Graphic Communication

Made Simple

The Made Simple series
has been created
especially for self-education
but can equally well
be used as
an aid to group study.
However complex the subject,
the reader is taken
step by step,
clearly and methodically,
through the course. Each volume
has been prepared by experts,
taking account of
modern educational requirements,
to ensure the most
effective way of
acquiring knowledge.

In the same series

Graphic Communication

Made Simple

Jack Whitehead

Made Simple Books
An imprint of Heinemann Professional Publishing Ltd
Halley Court, Jordan Hill, Oxford OX2 8EJ

OXFORD   LONDON   MELBOURNE   AUCKLAND   SINGAPORE
IBADAN   NAIROBI   GABORONE   KINGSTON

First published 1985
Reprinted (with revisions) 1989

British Library Cataloguing in Publication Data

Whitehead, Jack
    Graphic communication: made simple.—(Made
    simple books, ISSN 0265–0541)
    1. Engineering drawings
    I. Title   II. Series
    604.2   T379

ISBN 0 434 98602 X

Printed in Great Britain by
Richard Clay Ltd, Bungay, Suffolk

Preface

Technical Drawing has changed more drastically since about 1960 than in the whole century before. The dye-line and photocopying machines have destroyed the jobs of thousands of skilled copy-draughtsmen and women. At the same time, we see more drawings and pictures every day than the normal person a century ago saw in a whole year. Graphic Communication, the art of conveying information by all forms of drawing, has a long and interesting history, but has often been kept for the specialist. Today, the need to understand the secret languages of drawing, in wiring circuits, house plans, maintenance manuals and a host of text books, is becoming universal. Modern society needs to be literate, numerate and, if there is such a word, 'graphicate'.

The professional draughtsman or woman today, needs to be highly skilled and well trained in the broadest sense. Computer aided design, which is outlined in this book, is becoming more and more important in the design stage, hopefully saving months of testing and development. Production drawings tend to be made by highly skilled draughtsmen (the term 'draughtsman' is intended to include both sexes) who should be fully trained engineers understanding the process of manufacture thoroughly. Their drawings can be photocopied for a few pence, instantaneously and perfectly. A master copy, once checked, cannot be copied wrongly. The laborious copying by hand of master drawings has gone for good.

But graphics cannot be left to the professional. It must become a universal subject. Modern examinations reflect the recent changes. Graphic Communication covers a far wider field than the earlier Engineering Drawing type of syllabus.

This book will prepare students for GCSE examinations in Engineering Drawing and for the new Graphic Communication examinations which are bridging the gap between technology and art. Teachers and students on CPVE and BTEC courses will find this book a valuable resource. Some chapters, for example the one on perspective, will be of help to students in art colleges, while the chapter on surveying can give an extra dimension to Geography, Urban Studies or Archaeology courses.

In addition, the street corner photocopying machine has brought the chance of producing notices, advertisements or even pamphlets, to everyone. We all need to draw and to understand drawings. I hope that the general public will find the book of help and interest. It is time that Graphic Communication, one of the most important languages in the modern world, ceased to be a sort of medieval mystery and became part of the education of us all.

To Pat

To the Teacher

Every idea, visual aid and 'trick of the trade' in this book, has been used and tried out in a long experience of teaching from GCSE to Engineering Degree level. Some ideas have been picked up from others, some are original, but they all work in the classroom. No doubt you will have others or will improve on some of these. I have always found that having a new idea, or a new approach to an old topic, can bring zest to a lesson. Visual aids repay the effort of making them by silently teaching all the time and generating an atmosphere of study. I hope that you will find some of the ideas in the book helpful.

To the Student

A Guide to Study

Each part of this book deals with a different section of the drawing syllabus. There are so many ways of drawing that each has to be studied separately. Only later can one see them as a whole. At the same time, the book is arranged as a course, with the easier examples first and the more difficult ones later. Each type of drawing is developed so far, left for the time being, and a new type studied. Then, when all the drawing types have been introduced, the book deals with more advanced study of each. The page where the subject area continues, is given at the end of each relevant chapter. Thus anyone wishing to study one particular form of drawing, can follow it from chapter to chapter through the book. This should be particularly useful for revision.

Do not let me inhibit your choice, but the order of study I have used is based on a long experience and has always proved helpful to students.

Drawing Instruments

It is most important for you to obtain drawing instruments and start drawing from the beginning of the course. They need not be elaborate or expensive, but you must have them. Do not say, 'I will read the book and learn it and *then* do the drawing'. Draughtsmen and women are not made like that. If you want to learn to draw, you must draw. Anyone who knows it all in theory, but cannot put it on paper, is useless. You might as well be a silent radio announcer. If you want to be a draughtsman, you must spend hours and hours at a drawing board. I only hope you will enjoy it as much as I have done.

I should like to thank Simon Clements, Charles Thurgood, Mr W. A. Thornton and Geoffrey Whitehead for reading the text and making valuable suggestions. Any mistakes made in the book are, of course, my own.

Contents

List of Abbreviations

The list of abbreviations includes those which may appear in examination questions. Not all are used in the text.

A/F   across the flats
A ampere – unit of electricity
ASSY   assembly
BS   British Standards
CH HD   cheese head
CHAM   chamfered
CL   centre line
CRS   centres
CSK   countersunk
C'BORE   counterbore
CYL   cylinder
DRG   drawing
DIA   diameter
°   degree of angle
′   second of angle
FT or ′   foot measure
″   inch measure
GALV   galvanised
HEX   hexagonal
HEX HD   hexagon head
HP   horizontal plane
HT   horizontal trace
I/D   internal diameter
INT   internal
ISO   International Standards
   Organisation
Kg   kilogram
K   Kelvin – a degree of
   temperature with absolute zero
   as 0
LH   left hand
LVP   left vanishing point
M   metre
MATL   material
MAX   maximum
MIN   minimum

mm   millimetre
No.   number
ø   diameter
OD   outside diameter
PCD   pitch circle diameter
R or RAD   radius
RH   right hand
RD HD   round head
RVP   right vanishing point
SCR   screwed
SPHERE D   spherical diameter
SPHERE R   spherical radius
S'FACE   spot face
SP   station point
STD   standard
SH   sheet
T   taper
TPI   threads per inch
U'CUT   undercut
VP   vanishing point
     vertical plane
VT   vertical trace

1
Introduction

Drawing Instruments and Materials

Shops selling modern drawing equipment are Aladdin's caves, full of treasures, but bewildering. Pages 2–7 show some of the variey of instruments and materials on offer. You do not need them all. Gradually buy simple, durable instruments for pencil work. Ink drawing comes much later. Ink is not used for answering any examination, but is pleasant to try when you have done a good deal of pencil work.

You will need the following basic equipment:

A drawing board and T square
2H and 3H pencils for technical drawing and softer ones (B and 2B)
    for sketching
45° and 60° set squares
A drawing compass
A protractor, 180° or 360°
A scale in millimetres and inches
An eraser
Clips or pins to hold the paper on the board

Pencils

Pencil leads are made of graphite and clay. The more graphite, the blacker and softer the pencil. The more clay, the harder and lighter the lines will be. (See Fig. 1.1.)

6B–2B These are sketching pencils giving soft lines
B, HB and H These are normal writing pencils
2H and 3H These are the normal Technical Drawing pencils
4H–6H These give very fine lines but are too hard for our normal purposes

Drawing instruments (Fig. 1.2)

(*a*) A normal pencil of cedar or other even-grained wood.
(*b*) A modern fine-line pencil. The thin leads are protected by a metal tube which takes the wear.
(*c*) A clutch pencil, The lead is held by a spring-loaded, three-jaw collet.

2B

B

HB

F

H

2H

3H

4H

Fig. 1.1. Typical pencil lead shading showing the range of hardness and softness most draughtsmen use. Also available to 6B and 9H.

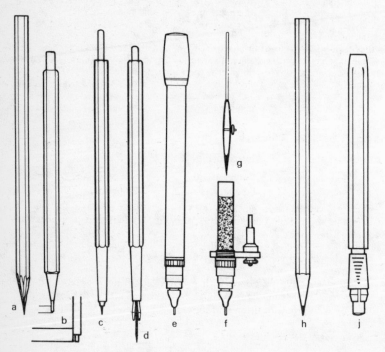

Fig. 1.2. Drawing instruments

(*d*) The spring has been pressed, moving the collet downwards and outwards, releasing the lead.

(*e*) A modern ink pen. This uses special ink which does not clog. The ink passes down a fine tube and gives a clean, even line. It is an excellent example of modern engineering and should be treated with the respect it deserves. *Note:* Thicknesses are illustrated in Fig. 1.3. Ink is not used for examinations but could be tried for course-work.

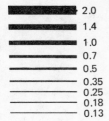

2.0
1.4
1.0
0.7
0.5
0.35
0.25
0.18
0.13

Fig. 1.3. Thickness of pen lines. Each pen draws one line thickness.

(*f*) The old spring pen, with two leaves, is easier to maintain than the new pens. It is cheaper and lasts longer, but it has to be used with skill.

(*g*) The pen with barrel removed to show the ink reservoir. It is fitted to a compass attachment.

(*h*) Blue 'non-print' or 'drop-out-blue' pencils are 'invisible' to photography and to some photocopiers. They are used to annotate drawings, etc. when one does not want the comments to be printed. The comments 'drop out'.

(*j*) The Staedler clutch eraser which has a round rubber. It is convenient for small work. (*Note:* The letter *i* is not used in labelling diagrams or making lists as it becomes confused with the number 1.)

Drawing Boards and T Squares

A variety of boards is illustrated in Fig. 1.4.

(*a*) A wooden board with tongued and grooved ends to keep it flat and prevent warping. The T square is made of even-grained fruit wood, beech or plastic.

(*b*) A board on a stand with a parallel ruler moving over top and bottom pulleys.

(*c*) The underside of a different type of parallel motion board showing four pulleys screwed flat to the board and the crossed wire which carries the rule.

(*d*) A lightweight, plastic board with grooved edges all round and a T square which slides in them. A quick release catch allows the square to be moved to a new position and locked.

Graphic Communication

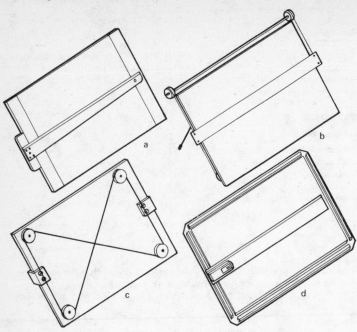

Fig. 1.4. Drawing boards

Set Squares (Fig. 1.5)

(*a*) and (*b*) Note that 45° and 60° set squares are made with different edges as illustrated at C.

(*c*) The square edge is suitable for pencil work when used either way up. Set squares with rulers along the edges are often bevelled with the measurements below, near the paper. These may be used, with bevel downwards, for ink work as the ink cannot run below the square and blot. The edge cut back on both sides can be used for ink work either way up.

(*d*) An adjustable square (set at almost 26° in Fig. 1.5) is rather a luxury, but it is a useful tool, especially for drawing sets of parallel lines. If you do buy one, make sure that the tightening nut is comfortably large and easy to use.

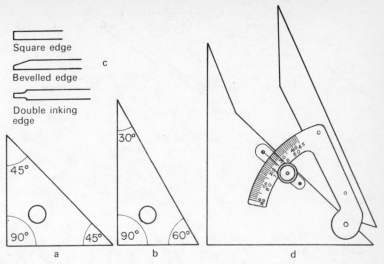

Fig. 1.5. Set squares

Compasses

These are available in a huge range (as shown in Fig. 1.6), but you can

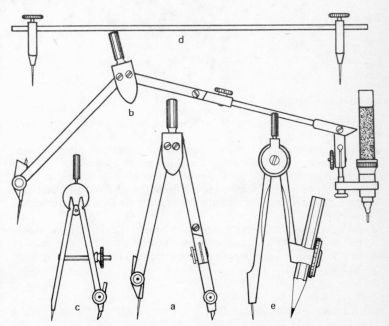

Fig. 1.6. A range of compasses – it looks rather like a convention of spider crabs

acquire them gradually. Ask for them for birthday presents. A detail draw-ing of a compass is shown in Fig. 14.16 but you will not be drawing to this level for some time.

Notes
(a) A pencil compass with an adjustable, double-ended needle and a hinged leg. Always use a lead one degree softer for your compass than for your pencil. The lines then become about equal in appearance.
(b) A compass fitted with a pen and an extension bar. The needle and pen are used vertically.
(c) A bow compass for small circles.
(d) A beam compass used as dividers. A pencil attachment could be used.
(e) The old pencil compass with which I passed my early exams, just to show that you do not need 'a million dollars worth of plumbing'.

Protractors

Fig. 1.7(a) shows a protractor divided into degrees and reading from 0° to 180° **in both directions**. Only a few numbers are shown to stress that you can reach 180° from left or right. Fig. 1.7(b) shows a protractor reading an angle of 117°.

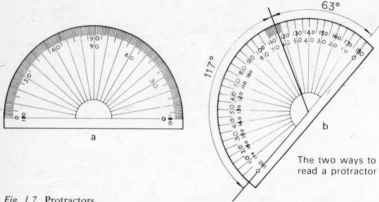

The two ways to read a protractor

Fig. 1.7. Protractors

Method Put one half of the base line on one arm of the angle and count round from the 0° mark on that arm. If you count in the wrong direction you will have the wrong answer of 63°.
Note: Airline pilots sometimes make the mistake of 'flying the reciprocal'. Using a 360° compass card, they fly in the opposite direction from that which they intend. Many an early airman's gravestone should read, 'He died because he read the compass card the wrong way round.' Always make sure which direction you are going.

Other Drawing Instruments

A wide variety of drawing aids is illustrated in Fig. 1.8. They include:

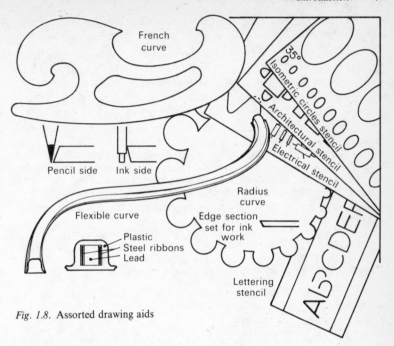

Fig. 1.8. Assorted drawing aids

French Curves

They are plastic curves for drawing smooth curves through fixed points. These are useful to skilled draughtsmen but can be dangerous for beginners. It is easy to draw beautiful, smooth curves which miss every point. Check your result carefully every time.

Flexible Curves

These are plastic strips with lead and spring steel inserts, which can be bent to any curve. They should be used with great care. Keep them flat: do not twist them.

Radius Curves

These are excellent for machine drawings. Choose them with pencil and ink sides. The ones with a knob on one side cannot be turned over and may not be suitable for ink work.

Stencils

There is an enormous variety of stencils available of circles, ellipses, electrical symbols, building symbols, etc. for engineers, electricians, and others. They are not cheap, so make sure you buy only the ones you need. Those with dimples underneath, lifting them off the paper, are better for ink.

Erasers (not illustrated)

Normal India rubbers or plastic erasers are quite satisfactory. Soft 'putty' rubbers, which can be kneaded into shape, are excellent. They lift the pencil lead from the paper instead of making crumbs. Keep the paper clean as you go.

A poor French student bought a single, stale loaf each day. The shopkeeper, a motherly woman, felt sorry for him as he grew thinner each week. It hurt her to think of this poor boy with nothing but his single, dry loaf, so one day, out of the kindness of her heart, she cut open the loaf and spread it thickly with butter. He returned next day and, to her surprise, attacked her with an axe. He was an architectural student. For weeks he had been preparing competition drawings for a new cathedral. They were to be handed in that very day. He had been using the bread to clean up his drawings.

Moral Keep grease, including the grease from your fingers, off the paper.

ISO (International Standards Organisation) Paper Sizes

The modern ISO 'A' paper sizes are calculated very ingeniously. The A0 sheet measures one square metre in area. Its sides are arranged in the proportions $1:\sqrt{2}$. Every time it is cut in half, the sides stay in the same proportion, You always have the same shape to draw on. How elegant.

Paper sizes

A sizes

A0 841 mm × 1,189 mm – 33.24″ × 47.24″ (one square metre)
A1 594 mm ×   841 mm – 23.62″ × 33″
A2 420 mm ×   594 mm – 16.5″ × 23.62″
A3 297 mm ×   420 mm – 11.81″ × 16.5″
A4 210 mm ×   297 mm – 8.25″ × 11.81″
A5 148 mm ×   210 mm – 5.9″ × 8.25″
A6 105 mm ×   148 mm – 4.12″ × 5.9″
A7   74 mm ×   105 mm – 2.95″ × 4.12″

See Fig. 1.9 for diagram showing the sizes.

B sizes

B1 1 metre × 700 mm   39.5″ × 27.75″
B2 700 mm × 500 mm   27.75″ × 19.75″
B3 500 mm × 350 mm   19.75″ × 13.87″
B4 350 mm × 250 mm   13.87″ × 9.87″

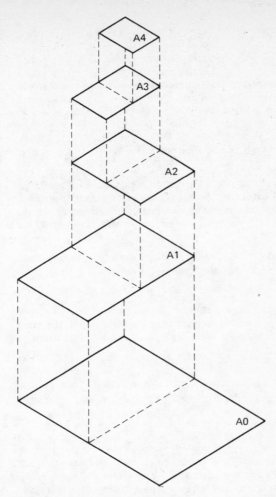

Fig. 1.9. Relationship of the 'A' sizes
Source: PD 7308: 1980 by permission of the British Standards Institution

The Weight of Papers

The weight is calculated in grams per square metre (gm^2).

Cartridge Paper This is the most common of all drawing papers – strong and firm. It should be a neutral white in colour and remain stable under all conditions. Fluorescent or blue–white papers, for example, can easily alter in colour in artificial light. The surface should not be too smooth and should have **size** – a gelatinous solution – worked into it to reduce ink penetration. Without size the ink 'feathers', spreading out on both sides. Size also prevents the surface from breaking up when erasing.

Detail Paper A thin, opaque paper, ideal for preparatory work. It is cheap, can often be traced through, but can be fragile.

Layout Paper A superior detail paper.

Tracing Papers Tracing paper fibres are very heavily beaten to achieve a high degree of transparency and to make the paper able to withstand repeated erasure. To test a tracing paper, draw an ink line, allow it to dry, and scrape it off with a sharp knife. Burnish with a pen top. Draw another ink line across it. If the ink spreads out in 'feathers', the paper is not good enough for ink work but may well do for pencil. Keep tracing papers dry. Water distorts them.

Tracing Film This is a plastic film with one or both sides matt coated to take drawing lines. Drawings inked on one side, can be updated by drawing on the underside. Tracing film is waterproof: it does not stretch or shrink. Keep it free of grease or the ink will not bite. It is available from 0.002″ to 0.01″ (50 microns or 0.05 mm to 25 microns or 0.025 mm) in thickness.

Tracing Cloth This is much less commonly used than it used to be, except where its lack of static is an advantage, as in laboratories, etc.

Special Papers Isometric and other grids – see the appropriate chapters.

Chalk Faced papers These have a matt chalk surface which gives a very black and white imprint suitable for paste-ups for offset-litho plate production or photocopy masters.

Art Papers These are coated on both sides with china clay and polished to a satin finish. Their fine, smooth surface ensures a faithful reproduction of colour work and their whiteness adds an extra brilliance to the colour. These are not drawing office papers but printers' papers.

Drawing Conventions

All lines should be uniformly black, dense and bold. Avoid lines which are too thin to be seen. The examiner may mark your paper at 2 a.m. with a cup of black coffee beside him. Do not expect him to see spider thin lines. The lines should be all in pencil or all in ink. (Do not use ink for exams).

Line Thicknesses

Two thicknesses are recommended: thin lines and thick ones. Thick lines should be two to three times the thickness of thin lines.

Types of line

Fig. 1.10 shows some of the types of line recommended by the British Standards Institution (BSI). **Always** use them or you will cause muddle and mistakes.

Type of line (Relative thicknesses)	Description of line	Application
(a) ————————	Thick, continuous	Visible outlines and edges
(b) ————————	Thin, continuous	Dimension and leader lines Projection lines Hatching Outlines of adjacent parts Outlines of revolved sections
(c) ∼∼∼∼∼∼	Thin, continuous irregular	Limits of partial views or sections when the line is not an axis
(d) _ _ _ _ _ _ _ _	Thin, short dashes	Hidden outlines and edges
(e) ——— - ———	Thin, chain	Centre lines Extreme positions of movable parts
(f) ——— - ——— - —	Chain (thick at ends and at changes of direction, thin elsewhere)	Cutting planes

Fig. 1.10. Types of line

Source: PD 7308: 1980 British Standards Institution

Lettering and Numerals

These should be clear, simple, well spaced and of a good size. All strokes should be black and of the same strength. No thicks and thins. Capitals only should be used.

Practice the letters and numbers regularly until good shapes become automatic. This is not the thing to learn up the night before the examination: it takes endless practice. Well shaped letters are the hallmark of the skilled draughtsman.

Do not underline. Use larger letters instead. (See Fig. 1.11.)

ABCDEFGHIJKLMN
OPQRSTUVWXYZ
1234567890

ABCDEFGHIJKLMNOP
QRSTUVWXYZ
1234567890

Fig. 1.11. Examples of letters and numerals

Note: Character height. The dimensions and notes should be not less than 3 mm tall. Titles and drawing numbers are normally larger.

Linear Dimensions

Linear measurements on drawings are normally in millimetres today, but many countries still use feet and inches. A draughtsman may need to draw **and think**, in both. The thinking is the more difficult. Figs. 1.12, 1.13 and 1.14 show a number of important conventions which must be learnt carefully. Notice the following points: the decimal marker is a full point placed on the base line, e.g. 27.6; print 35 not 35.0; *but* print a zero before a size less than unity, e.g. 0.5.

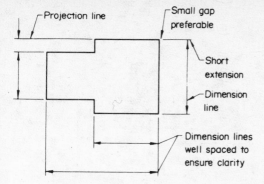

Projection and dimension lines

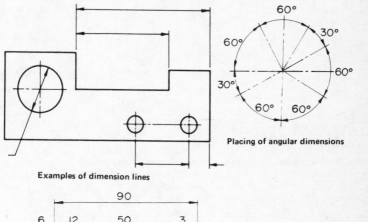

Examples of dimension lines

Placing of angular dimensions

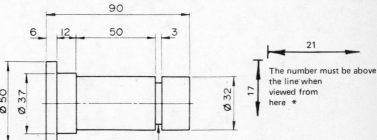

Placing of linear dimensions: larger dimensions placed
outside smaller dimensions

Fig. 1.12. Linear dimensions

Source: PD 7308: 1980

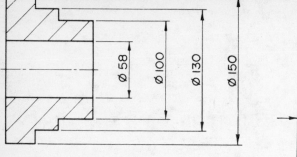

Use of the diameter symbol ⌀

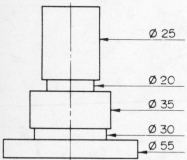

Dimensions applied to features by leaders

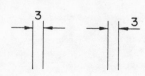

Dimensioning of small features

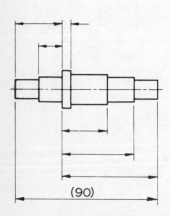

**Overall length added
as an auxiliary dimension**

Fig. 1.13.

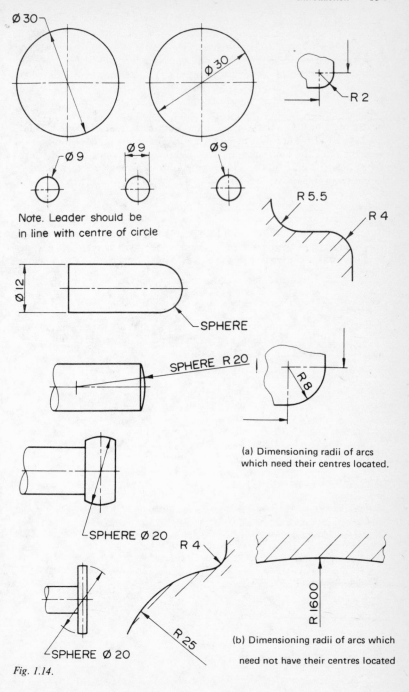

Note. Leader should be in line with centre of circle

(a) Dimensioning radii of arcs which need their centres located.

(b) Dimensioning radii of arcs which need not have their centres located

Fig. 1.14.

The Arrangement of Measurements

It is not easy to arrange measurements so that they can be read easily. A drawing must be clear and must not be misunderstood. Therefore you *must* use the standard drawing conventions. You are not free to do what you like. If you do, you may ruin your factory by causing those who use your drawings to produce unsaleable scrap.

Place numbers near the middle of the dimension line and clear of it. The figures **must be above the dimension line when seen from the bottom right-hand corner of the drawing.** (See Fig. 1.12.) The arrow heads and numbers must be dark: arrow shafts must be light. Practice drawing neat, clear arrow heads. Notice all the different methods of arranging measurements, as shown in Figs. 1.12, 1.13 and 1.14.

Circles

The diameter sign is a circle with a line across it (like Saturn) and is always placed **in front** of the numeral. Try to move the measurements well away from the drawing whenever possible. Never put a dimension along a centre line.

Angles

Try to place the figures horizontally. They should not be at odd angles. (See Fig. 1.12.) Good placing of measurements is an art worth cultivating and can greatly enhance a drawing.

2
Plane Geometry I

Plane Geometry is the geometry of flat shapes. They have length and breadth but no thickness. A **plane** is a flat surface. Mathematicians have worked out many sorts of geometry, each with its own rules. The rule of Plane Geometry is that everything must lie flat. A square on the surface of the world has length and breadth but it will not lie flat so it comes into Spherical Geometry, not Plane. This book does not deal with Spherical Geometry. You will deal with that if you become an airline pilot.

A Visual Aid on Lines

A draughtsman's work consists largely of lines. There is a special vocabulary about lines which we must all know. These words will be used throughout the book and must be learnt by heart. Your teacher/lecturer may display this visual aid in an enlarged form, or you may study Fig. 2.1 carefully.

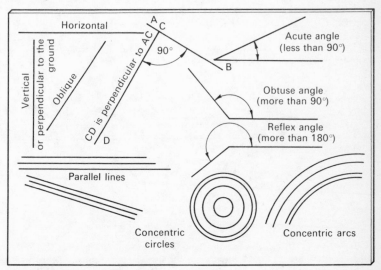

Fig. 2.1. Types of line: A visual aid

A Visual Aid on Plane Figures

The visual aid (see Fig. 2.2) may be made in sheet plastic or painted hardboard with corner holes for hanging. Labels should be large. The names must be learnt by heart and also the spellings. These words are in constant use in any book on Graphics. Notice the difference between **regular** and **irregular** figures. Regular figures have all sides and angles equal. Irregular ones do not.

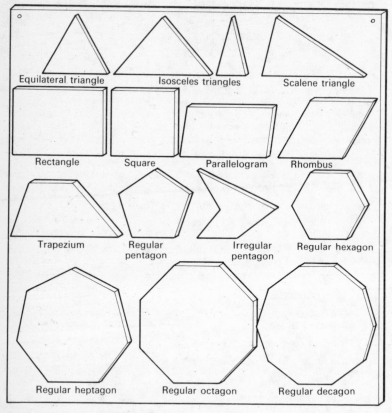

Fig. 2.2. Plane figures: A visual aid

Bisection of Lines and Angles

The Bisection of a Line

A section is a cut. Bisection is cutting into two equal parts.

Method Draw a line AB. Set the compass to more than half AB. With centre A draw an arc above and below the line. Repeat from centre B with the **same** radius to cut at C and D. Join C and D. This line **bisects** AB. The line AB can be at any angle. Lines can be divided into two; four; eight; etc. by repeated division. (See Fig. 2.3.)

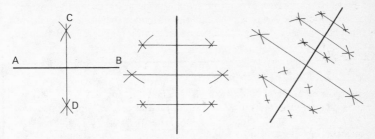

Fig. 2.3. The bisection of lines

The Bisection of Angles

To bisect the angle ABC. (See Fig. 2.4.) Draw an arc with centre B and any radius, to cut the angle at D and E. Draw arcs of *equal* radius with centres D and E to cross at F. Join FB. The line FB bisects the angle ABC.

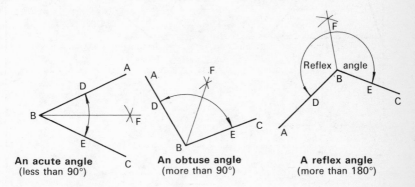

An acute angle (less than 90°)   **An obtuse angle** (more than 90°)   **A reflex angle** (more than 180°)

Fig. 2.4. The bisection of angles

The Angles of a Circle

A circle is one complete turn which we divide into 360°. The sign for a degree is °. The story goes that the Ancients thought that the Sun took 360 days to go round the Earth. Each day marked out a degree. In fact the Earth takes about $365\frac{1}{4}$ days to circle the Sun. The choice of 360° in a circle is very convenient because 360 can be divided by 2, 3, 4, 5, 6, 8, 10, 12, 15, 18, 20, 24, 30, 36, 40, 60, 72, 90, 120 and 180 (see Fig. 2.5). A truly amazing number. This is why we have retained 360° for the sub-division of a circle.

360° ÷ by	2	3	4	5	6	8	9	10	12	15	16	18
equals	180°	120°	90°	72°	60°	45°	40°	36°	30°	24°	22½°	20°

360° ÷ by	20	24	30	36	40	45	60	72	90	120	180
equals	18°	15°	12°	10°	9°	8°	6°	5°	4°	3°	2°

Fig. 2.5. These are the divisions of 360°. No number below 360 has so many factors or would be so useful.

To Construct Angles with Compasses

To construct an angle of 45°. Bisect 90° as shown in Fig. 2.7.

To construct an angle of 60°. The radius of a circle can be stepped out exactly six times round a circle (see Fig. 2.8). This gives an angle of 60°.

To construct an angle of 30°. Bisect an angle of 60°. Bisect again for an angle of 15°. (See Fig. 2.9.) It is impossible to tri-sect a line with a compass.

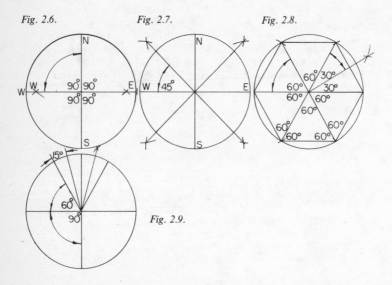

Fig. 2.6.   *Fig. 2.7.*   *Fig. 2.8.*

Fig. 2.9.

Parts of a circle

These names are very important and must be learnt by heart. Again, they will be used throughout the book. (See Fig. 2.10.)

Fig. 2.10. The parts of a circle

Diameter A straight line through the centre of a circle from the circumference on one side to the circumference on the other.
Radius A straight line from the centre of a circle to any point on the circumference.
Circumference The outside line of a circle *or* the distance round a circle.
Arc Any part of the circumference.
Chord A straight line touching the circumference at both ends.
Tangent A straight line which just touches a circle.
Sector A wedge shaped part of a circle bounded by two radii and an arc.
Segment A section bounded by a chord and an arc.

The Circumference of a circle (See Fig. 2.11.)

The circumference of a circle equals the diameter $\times$ π (pronounced 'pie').
The diameter would wrap round the circumference π times.

$$\pi = \text{approximately } 3\tfrac{1}{7} \text{ or } \tfrac{22}{7} \text{ or } 3.14$$

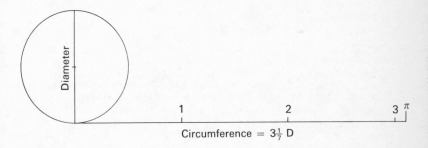

Fig. 2.11. The circumference of a circle

The Area of a Circle

Example Given that the radius of a circle = 6 mm.

$$\begin{aligned}
\text{Area} &= \pi r^2 \\
&= 3\tfrac{1}{7} \times 6 \times 6 \\
&= \tfrac{22}{7} \times 36 \\
&= \tfrac{792}{7} \\
&= 113\tfrac{1}{7} \text{ sq mm}
\end{aligned}$$

Fig. 2.12 shows the area of a circle laid out in squares. The square continuing the circle has been divided into 100 small squares and the area counted as follows:

$$\begin{aligned}
71 \text{ whole squares} &= 71 \\
4 \text{ three-quarter squares} &= 3 \text{ approximately} \\
4 \text{ half squares} &= 2 \quad \text{,,} \\
4 \text{ quarter squares} &= 1 \quad \text{,,}
\end{aligned}$$

$$\text{Total} \quad 77 \text{ small squares}$$

This is slightly smaller than it should be.

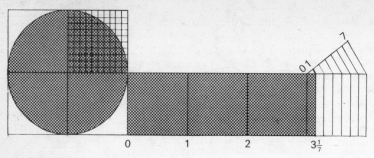

Fig. 2.12. The area of a circle

Questions

Print the name of each line and angle you draw.

1 Draw a vertical line 39 mm long and bisect it.
2 Draw an oblique line 68 mm long and bisect it.
3 Draw any acute angle and bisect it.
4 Draw an obtuse angle and bisect it.
5 Draw a reflex angle and bisect it. Then bisect one half and bisect the quarter.

To divide a circle into four equal parts (See Fig. 2.6.)

Draw a circle 60 mm in diameter. Draw a vertical diameter NS and bisect it to give EW. We have divided the circle into four equal parts and constructed 90° angles.

Angles can be constructed by using compasses or with protractors.

Set Squares

To Construct Angles using Set Squares (see Fig. 2.13) ***and a T square***

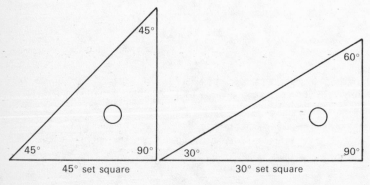

45° set square   30° set square

Fig. 2.13. Set squares

Fig. 2.14 shows the constructions of angles of 75°, 105°, etc.

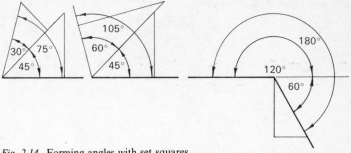

Fig. 2.14. Forming angles with set squares

Questions (continued)

6 Draw Fig. 2.6 using a 60 mm diameter circle.
7 Bisect the result to give an angle of 45°.
8 Draw a 35 mm radius circle and copy Fig. 2.8.
9 Bisect and re-bisect to divide a 64 mm diameter circle into 15° sections.
10 Draw a 100 mm diameter circle and copy Fig. 2.10 with good lettering.
11 Using set squares, construct the following angles: 75°, 120°, 135°, 210°.

Protractors

Protractors are made as semicircles measuring 180° or as complete circles measuring 360°. All protractors *measure in both directions*.

To measure an Angle ABC

Place the protractor with its centre over the apex of the angle (B) and the baseline (the 0° line of the protractor) on BC. Read along the scale which starts at 0 on BC. Do not muddle the scales. Always count from 0°. Fig. 2.15 shows the same angle measured using the different scales.

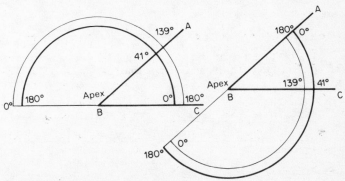

Fig. 2.15. Measuring angles with protractors

To Draw a Perpendicular from a Point to a Line

(See Fig. 2.16.) With centre P draw an arc to cut line AB at C and D. From C and D draw equal arcs to cross at E.

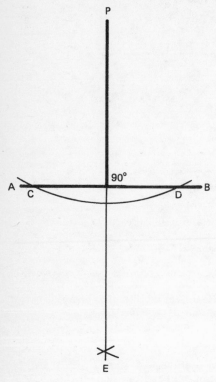

Fig. 2.16. To draw a perpendicular from a point to a line

To Draw a Perpendicular from a Particular Point in a Line

(See Fig. 2.17.) Let C be a point on a line. With centre C draw any arc DE. Mark off arcs equal to CD at F and G. Draw equal arcs to cross at F. FC is perpendicular to AB.

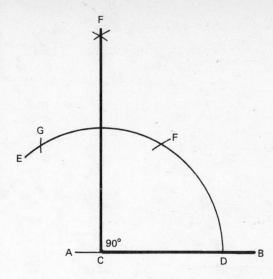

Fig. 2.17. To draw a perpendicular from a line to a point

To Draw Angles by Repeated Bisection

Fig. 2.18 shows the bisection and rebisection of a right angle to give $45°$, $22\frac{1}{2}°$, $67\frac{1}{2}°$.

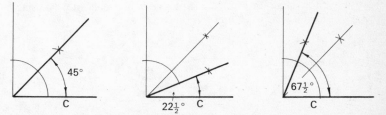

Fig. 2.18. Bisection of angles with a compass

The Construction of a Magnetic Compass Card by Bisection

This is built up as in Fig. 2.19 but repeatedly bisected. South South East ($157\frac{1}{2}°$) is the reciprocal of North North West ($337\frac{1}{2}°$).

Magnetic Compass

The magnetic compass was one of the great liberators of mankind, allowing us to travel and explore. Early maps are networks of lines radiating from different ports and cities. We expect maps to be covered with longitude

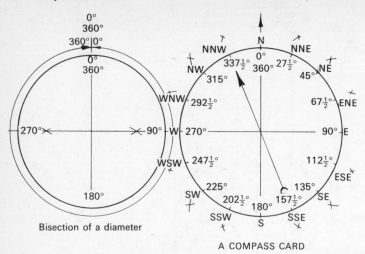

Fig. 2.19. A compass card by repeated bisection

Note: All letters and numbers must be horizontal

and latitude lines, but these came centuries later. Early merchants knew that if you travelled on a compass bearing for a certain number of days, you would reach your target, so early maps were compass bearings and lines of different lengths of travel, drawn to a scale.

A Simple Compass Course

Plot the course shown in Fig. 2.20 as a navigator would plot it, using instruments and a scale of 20 mm = 1 mile.

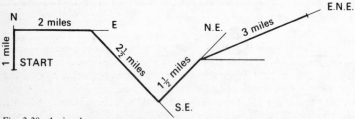

Fig. 2.20. A simple compass course

1 Mark the starting point. Draw a line to the North. Measure 20 mm (1 mile) along it. Then 2 miles East, 2½ miles SE, 1½ miles NE, 3 miles ENE. How many miles are you from the starting point?

Now plot these courses in the same manner, to the same scale.

2 Two miles South, 3 miles NW, 2½ miles NE, 1½ miles ENE. How far are you from base?

3 Four miles North, 5 miles West, 3 miles SW, 10 miles ENE. How far are you from base?

Answers 1. 7.8 miles 2. About 2.65 miles 3. About 6.05 miles.

Division of a Line

To Divide a Line into Three Equal Parts (see Fig. 2.21)

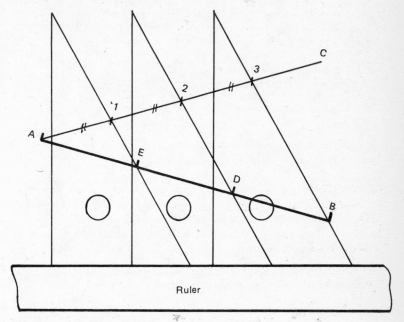

Fig. 2.21. The division of a line

Method Draw a line AB and *any* line AC at *any* angle. Measure three equal distances from A along the line AC. Join 3B. Place one edge of a set square along the line 3B. Place a straight edge to touch another edge of the square. Keep the ruler still and slide the set square along until it touches 2. Draw 2D. Repeat for 1E. Then AE = ED = DB. A1E, A2D and A3B are **similar triangles**. They have the same angles and are in the same proportions (see Fig. 2.22).

To Divide a Line in Given Proportions

If you want to divide a line 50 mm long in the proportions 3:2 by calculation, you:

(*a*) add the number of parts (2 + 3 = 5)
(*b*) divide 50 by 5 = 10
(*c*) multiply 2 × 10 and 3 × 10. This gives the two parts, 20 and 30.

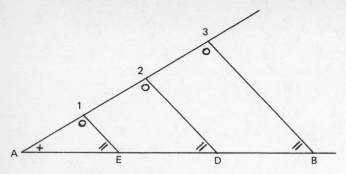

Fig. 2.22. Similar triangles A1E, A2D and A3B

Angle A1E = A2D = A3B
Angle 1EA = 2DA = 3BA
Angle 3AE is common
Lengths AE = ED = DB
∴ AE = ED = DB

We can also draw the solution and, when the figures are complicated, this is often quicker.

Method (see Fig. 2.23). Draw a line AB equal to 50 mm. Draw a line AC at *any* angle. Mark five equal distances along AC. Join 5B. Draw 3D parallel to 5B. Then AD, DB are in the proportions 3:2

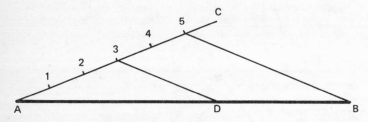

Fig. 2.23. To divide a line in given proportions

To Divide a Line 93 mm Long in the Proportions 3:4

Draw a line AB 93 mm long (see Fig. 2.24). Draw a line AC at any angle. Mark seven equal divisions along AC from A. Join 7B. Draw 3D parallel to 7B. Then AD and DB are in the proportions 3:4.

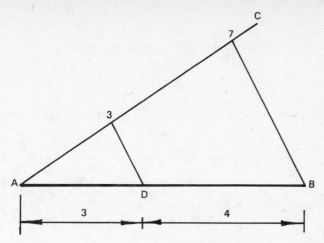

Fig. 2.24. To divide a line 93 mm long in the proportions 3:4

To Draw a Triangle with a Perimeter of 113 mm and Sides in the proportions of 3:4:5

Draw a line OP equal to 113 mm (see Fig. 2.25). Divide OP into twelve equal parts because 3 + 4 + 5 = 12. Then AB is the side of the triangle having 5 parts. With centre A and radius OA, make an arc OC. With centre B and radius BP draw an arc PC to cut arc OC. Draw AC and BC. Then ABC is the required triangle. Measure angle ABC. What is the 3:4:5 triangle used for?

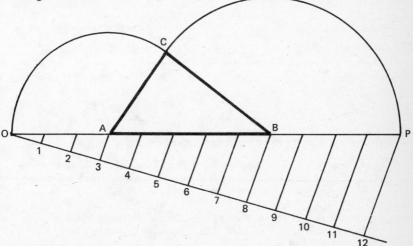

Fig. 2.25. To draw a triangle with a perimeter of 113 mm and sides in the proportions 3:4:5

Questions

Draw all the above. Then solve the following problems:

1 Divide a line 121 mm long into five equal parts.
2 Divide a line 99 mm long into seven equal parts.
3 Divide a line 87 mm long in the proportions 2:5.
4 Divide a line 107 mm long in the proportions 3:5.
5 Draw a triangle, perimeter 149 mm and its sides in the proportions 2:4:5. Measure all the angles.

Plain Scales

A **plain scale** is like a ruler, but it is numbered from 0 so that the units are on the right and the fractions are on the left. Fig. 2.26 shows a plain scale measuring up to 6 metres. The left hand metre has been divided into tenths of a metre. Arrow (*a*) measures 5.8 metres. What does arrow (*b*) measure?

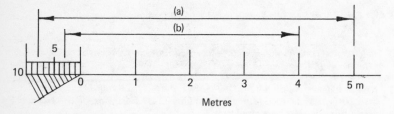

Fig. 2.26. A plain scale

Construct the right angled triangle in Fig. 2.27 and find the length of the

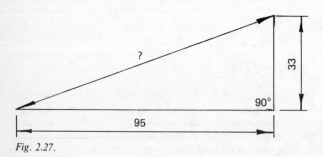

Fig. 2.27.

hypotenuse by using the scale Fig. 2.27 and find the length of the hypotenuse by using the scale Fig. 2.26. (The hypotenuse is the longest side and is opposite the largest angle.) Fig. 2.28 shows a plain scale measuring up to 4 km at a scale of 1:100,000 (*Note:* There are 100,000 centimetres in a kilometre).

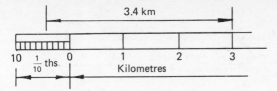

Fig. 2.28. A plain scale where 1 cm = 1 km

Questions

1 Using the km scale, construct the compass course 3.6 km North, 6.3 km West, 2.9 km South, 8.2 km East. How far are you from the start?
2 Draw a plain scale in metres and tenths of a metre where 2 cm equals 1 m, to measure up to 6 metres.

   Method Draw this step by step. The total length of the scale will be 6 × 2 cm = 12 cm. Draw a scale 12 cm long, divide the left hand section into 10 equal parts and number correctly from 0 in both directions. Draw very exactly. The scale is 2 cm = 1 m or 1 : 50 (there are 50 lots of 2 cm in 1 m). Always put the size you have drawn **first**.
3 Draw an **enlarging** plain scale to measure to 1 m where 20 mm = 0.1 mm. This could be used for drawing small fossils measured under a microscope. The scale is to measure up to 1 mm. (The total length of the scale will be 200 mm.) Print the scale ? = 1.
Answer 1:200
4 Construct a scale to measure up to 4 km in units of 100 metres. Scale 40 mm = 1 km. Print the scale as a proportion (1:?).

Diagonal Scales

Diagonal scales are ingenious ways of drawing rules which can be used for very accurate measurement. They can measure directly up to ten times as accurately as a plain scale.

   They are simpler than they look.

A Diagonal Scale Drawn in Stages (see Fig. 2.29(a), (b), (c))

The bottom line is drawn first (see Fig. 2.29(a)) and looks like a plain scale e.g. (a) = 1.7. Draw vertical lines from each point. Draw horizontal lines at equal distances apart parallel to the base line. Number the lines at the right hand end from the bottom to the top. Stagger the numbers if necessary. We now have a series of plain scales, one above the other (see Fig. 2.29(b)). Draw the diagonal from bottom 0 to top 1. Repeat this slope all the way along. Number the scale carefully as shown. (See Fig. 2.29(c).) Consider one diagonal. The 1.4 diagonal goes from the bottom at 1.4 to the top at 1.5. The horizontals cut it into 10 equal pieces. These are tenths of tenths = hundredths.

Lengths are measured by counting the units. Then the tenths and hundredths are counted in an *anticlockwise direction*. 1 unit, 4 tenths and six hundredths = 1.46.

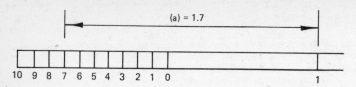

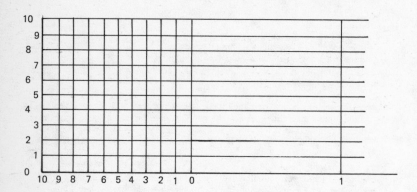

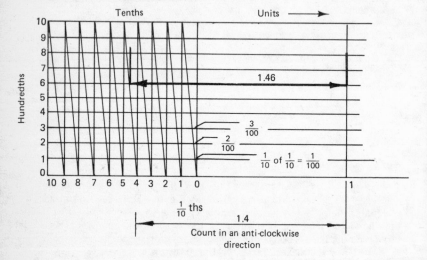

Fig. 2.29 A diagonal scale drawn in stages.

 (*a*) A plain scale
 (*b*) A pile of plain scales
 (*c*) A diagonal scale

Note: 1.46 is between 1.40 and 1.50 so measure along the 6th hundredths line as far as the diagonal line between 0.5 and 0.6

Questions

1 Draw a diagonal scale in kilometres and hundredths of a kilometre to measure up to 4 km. Scale 5 cm = 1 km or 1:20,000 as there are 20,000 lots of 5 cm in 1 km.
2 Draw a diagonal scale in metres and hundredths of a metre to measure up to 4 m at a scale of 1:50. Use it to draw a rectangle 3.21 m by 2.93 m and measure the diagonal.

Triangles

A triangle has three straight sides. The angles of a triangle always add to 180°.

Construct the Triangles in Fig. 2.30

Measure all angles with a protractor and fill them in. Add measurements in millimetres).

1 **An equilateral triangle with sides of 65 mm.** Equilateral triangles have **all sides and**

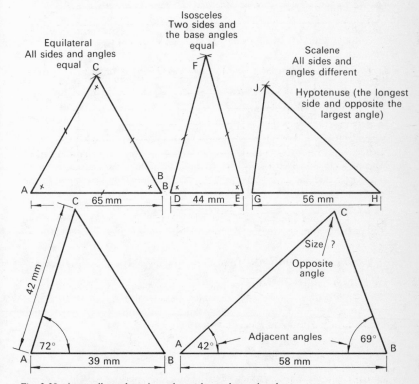

Fig. 2.30. An equilateral, an isosceles and a scalene triangle
Fig. 2.31. To draw a triangle given two sides and the included angle
Fig. 2.32. To draw a triangle given one side and two adjacent angles

all angles equal. Draw a base AB equal to 65 mm. Strike arcs of 65 mm radius from the centres A and B to cross at C. ABC is the required equilateral triangle.

2 **An isosceles triangle base 44 mm and two equal sides of 70 mm.** An isosceles triangle **has two sides and the base angles equal.** Draw a base DE 44 mm long. Draw arcs of 70 mm radius from the centres D and E to cross at F. DEF is the isosceles triangle.

3 **A scalene triangle with sides 72 mm, 56 mm and 43 mm.** A scalene triangle has **all sides and angles different.** Draw a base line GH 56 mm long. With the centre G strike an arc of 43 mm radius. With the centre H strike an arc of 72 mm radius to cut it at J. GHJ is the required scalene triangle.

Measure all angles and check that they add up to 180°.

To Draw a Triangle given Two Sides and the Included Angle

To draw a triangle with one side 39 mm, the next 42 mm and the included angle 72°. (See Fig. 2.31). Draw a line AB 39 mm long. Measure an angle of 72° at A. Measure a line 42 mm from A to C. Join BC, measure it and measure the other angles.

To Draw a Triangle given One Side and the Two Adjacent Angles

(See Fig. 2.32.) Draw a line AB 58 mm long. Draw an angle BAC of 42°. Draw an angle ABC equal to 69°. Where the two lines cross will be C.

To Draw a Triangle given Two Sides and the Opposite Angle

Draw a triangle ABC in which AB = 83 mm, AC = 72 mm and angle ACB = 51°. (See Fig. 2.33.)

Method Draw a line AB 83 mm long. Bisect AB with the line EF. Construct an angle BAD = 51°. Draw an angle DAP at 90° to cut EF at O. With centre O and radius OA, draw a circle. Draw an arc, centre A and radius 72 mm to cut the circle at C. Join AC and BC. This construction is based on the fact that the angle between a chord and a tangent at one end of the chord, is equal to the angle subtended (supported by) the chord in the opposite segment of the circle. (Angle BAD = the angle at C or G or H etc.) C is the only point that has AC equal to 72 mm and therefore is the only possible apex for us.

To Construct a Right Angled Triangle by Making a 3:4:5 sided Triangle

Any triangle with sides in the proportion 3:4:5 must contain a right angle. The Egyptians, and later many other builders, set out right angles by means of a rope knotted at twelve equal distances. This was pegged out to give the right angle. It makes a useful **visual aid** (see Fig. 2.34).

To Construct a Right Angled Triangle with a Given Perimeter

(See Fig. 2.35.) Let the perimeter be 100 mm. Draw a line AB equal to 100 mm. Draw a line AC at *any* angle and mark off twelve equal lengths. Join 12B. Draw lines 3D and 7E parallel to 12B. Then AD, DE and EB are in

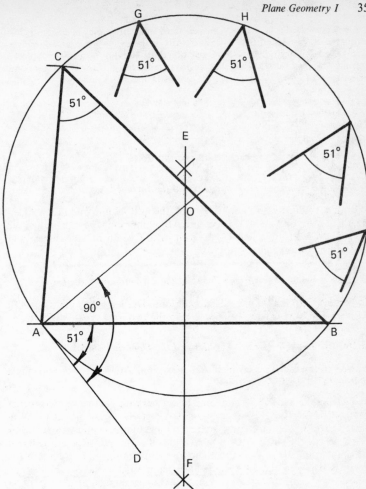

Fig. 2.33. To draw a triangle given two sides and the opposite angle

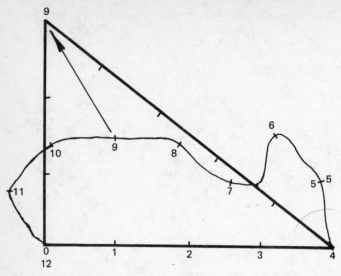

Fig. 2.34. A visual aid to right angled triangles

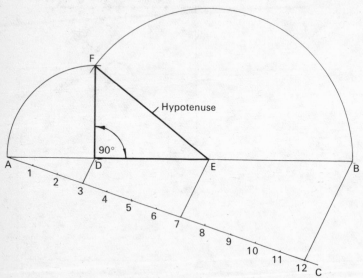

Fig. 2.35. To construct a right angled triangle with a given perimeter

the proportions 3:4:5. With the centre D and the radius DA, draw an arc. With the centre E and the radius EB, draw an arc to cut the first arc at F. DEF is a triangle with 3:4:5 sides and the angle DFE is a right angle.

Visual Aid to show Right Angles in a Semicircle

Make a semicircular frame in thick wire (as shown in Fig. 2.36). Slip on a small ring and complete the joint on the diameter. String up with elastic cord and fasten to a board at the diameter only. Alternatively, use curtain rail and hook-sliders.

Questions on Triangles

1 Draw a triangle with sides 49 mm, 72 mm and 72 mm and name it.
2 Draw a triangle with sides 69 mm, 48 mm and an included angle of 77°.
3 Draw a triangle with a base of 59 mm and base angles of 46° and 66°. Measure the vertical height.
4 Try to draw a triangle with sides 96 mm, 33 mm and 45 mm. Why is this impossible? Write out a rule about the lengths of sides of a triangle.
5 Draw a triangle with a base of 76 mm, another side 52 mm and an opposite angle of 63°. How many answers are there to this problem?
6 Draw the largest possible equilateral triangle in a circle with a diameter of 70 mm. (Draw a diameter and set off 30° each side.)
7 Construct a right angled triangle of perimeter 117 mm. Measure the hypotenuse. (The hypotenuse is the longest side.)
8 Draw a triangle with sides 49 mm and 67 mm and the included angle of 107°. Measure the vertical height, which will fall outside the base.

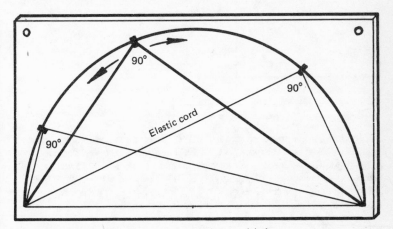

Fig. 2.36. A visual aid to show right angles in a semicircle

The Construction of Rectangles

Rectangles are quadrilaterals with opposite sides equal and parallel and all angles right angles. Fig. 2.37 shows the construction of a rectangle. Draw the base AB. Erect a perpendicular at A using a compass. Measure the height AC. Draw an arc with a radius of AB from the centre C. Draw an arc with a radius of AC from the centre B to cut the first arc at D. Join CD and BD.

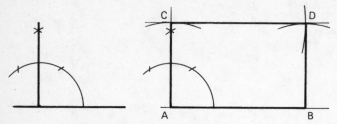

Fig. 2.37. The construction of a rectangle

The area of a rectangle = length × height (see Fig. 2.38)
The area of a triangle = length × half the vertical height

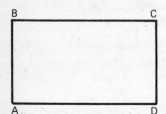

Fig. 2.38.

To Find the area of Any Triangle ABC

Let the line AC be the base (see Fig. 2.39). Draw a perpendicular from the apex B to the base AC to cut AC at D. Then BD is the vertical height of the triangle. Draw a rectangle of length BC and height BD. Then triangle ABD has half the area of rectangle AEBD. Triangle DBC is half DBFC. Therefore ABC has half the area of rectangle AEFC. The areas of many figures can be found by cutting them up into triangles.

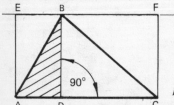

Fig. 2.39. To find the area of any triangle

Squares

To Construct a Square Given the Length of Side

Draw a square of 65 mm side. Construct as for a rectangle (see Fig. 2.40).

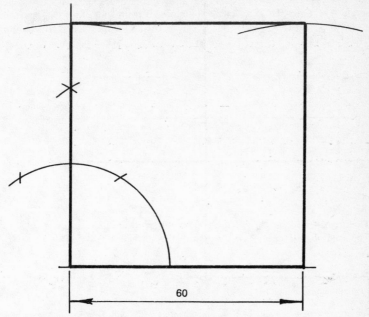

60

Fig. 2.40. To construct a square given the length of side

To Construct a Square Given the Diagonals

Draw one diagonal AB 73 mm long. Draw the sides at 45° angles to the diagonal. Measure the length of the side. (See Fig. 2.41.)

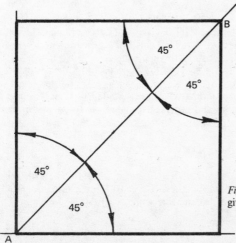

Fig. 2.41. To construct a square given the diagonals

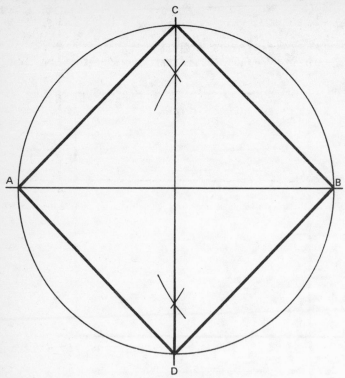

Fig. 2.42. To draw the biggest square that will fit into a circle

To Draw the Biggest Square that will Fit into a Circle

Draw a circle of 70 mm diameter. Draw a diameter AB. Bisect AB to give the diagonal CD. ABCD is the largest square.(See Fig. 2.42.)

Rectangles

To Construct a Rectangle Given One Side 80 mm Long and a Diagonal 87 mm Long

Draw the side AB 80 mm long. Construct a second side BC at right angles to AB. With a radius of 87 mm and centre A draw an arc to cut BC at D. Construct the fourth corner by striking arcs. (See Fig. 2.43.)

Alternative Construction of Rectangles

Draw a circle of an 87 mm diameter. Draw the diameter AC. With the centre A and a radius of 80 mm, cut the circle at B. Repeat for corner D. (See Fig. 2.44.) This method is not suitable when AB must be horizontal.

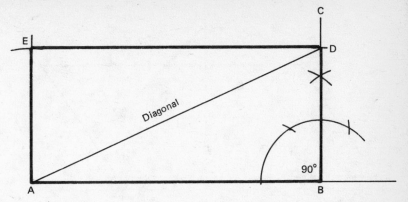

Fig. 2.43. To construct a rectangle given one side 80 mm long and a diagonal 87 mm long

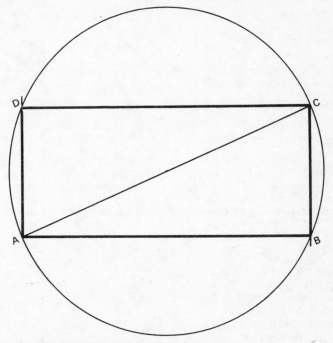

Fig. 2.44. Alternative construction of rectangles

Parallelograms

Parallelograms have opposite sides equal and parallel, but the corners are not right angles.

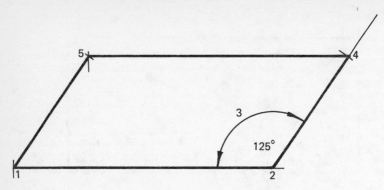

Fig. 2.45. To draw a parallelogram with sides of 80 mm and 40 mm and an angle of 55°

To Draw a Parallelogram with Sides 80 mm and 40 mm and an Angle of 55°

Draw in number order shown in Fig. 2.45.

To Construct a Parallelogram Given the Length of Sides as 63 mm and 42 mm and a diagonal 97 mm long

Construct in number order as shown in Fig. 2.46.

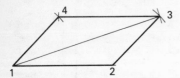

Fig. 2.46. To construct a parallelogram given the length of sides as 63 mm and 42 mm and a diagonal 97 mm long

The Angles and Lengths in a Parallelogram (see Fig. 2.47)

Key:
X = opposite angles
O = opposite angles
Z = alternate angles
Y = alternate angles (also P and Q)
/   = equal lengths and parallel
// = equal lengths and parallel

These names should be learnt as they may be used frequently.

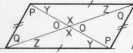

Fig. 2.47. The angles and lengths in a parallelogram

Rhombus

A **rhombus** is a parallelogram with all sides equal.

To Construct a Rhombus with Sides 60 mm Long and an Included Angle of 120°

Construct in number order as shown in Fig. 2.48. Measure the short diagonal. Why is it this size?

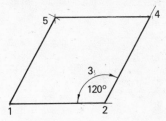

Fig. 2.48. To construct a rhombus with sides 60 mm long and an included angle of 120°

Trapeziums

A **trapezium** is a quadrilateral with two of its sides parallel.

To Construct a Trapezium

Given:
1 Lengths of sides 96 mm, 49 mm, 27 mm and 68 mm.
2 Order of sides – as above.
3 The vertical height 46 mm.

Construct in number order as shown in Fig. 2.49(*a*).

Given:
1 The lengths of sides.
2 The order of sides.
3 One diagonal.

Construct in number order as shown in Fig. 2.49(*b*) there are two solutions, the mirror images of each other.

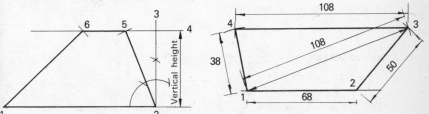

Fig. 2.49. To construct a trapezium. ((*a*) and (*b*))

Regular polygons

The regular figures and regular solids, which we shall study later, were greatly admired by the Greeks because they thought 'perfect' shapes were more noble than ordinary ones.

A **Perfect Figure** has sides of equal length and all angles equal. They are as follows:

Name	No. of sides	Angles
Equilateral triangles	three sides	60° (learn)
Squares	four sides	90° (learn)
Regular pentagons	five sides	108° (learn)
Regular hexagons	six sides	120° (learn)
Regular heptagons	seven sides	not an exact degree
Regular octagons	eight sides	135° (learn)
Regular nonagons	nine sides	odd
Regular decagons	ten sides	144°
Regular dodecagons	twelve sides	150°
Circles	An infinity of sides	No corners

The plans of many buildings start from one or more of these figures to give a unity and 'shape' to the space. Studying plans of buildings to find their underlying geometrical form, can be very enjoyable. The geometrical plan of Florence Cathedral (see Fig. 2.50) decided the whole arrangement of pillars, roof shapes and the glorious octagonal dome.

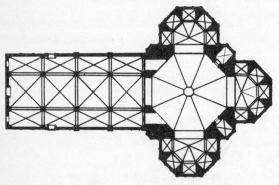

Fig. 2.50. Plan of Florence Cathedral. Based on an octagon, squares and rectangles – an example of practical geometry.

The Golden Mean

Many Greek and Roman buildings have windows made to a proportion called the 'Golden Mean'. This is a very pleasing shape of about 1:1.6. Other architects copied the proportion so that one can find it in many eighteenth-century buildings in Europe and America. Whole books have

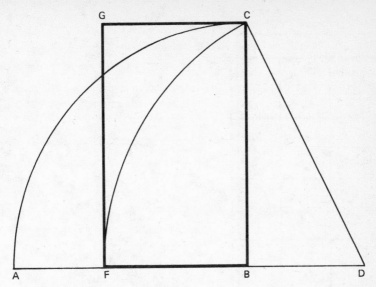

Fig. 2.51. To draw a rectangle with sides in the proportion of the Golden Mean

been written on architectural proportion. Some are of little real value but the Golden Mean is always a useful working tool.

To Draw a Rectangle with Sides in the Proportion of the Golden Mean

(See Fig. 2.51.) Draw a line AB equal to the long side of the rectangle. Erect BC at right angles with AB. Extend AB to D so that BD equals ½AB. With the centre D and the radius DC, draw an arc to cut AB at F. Draw rectangle FGCB.

Fig. 2.52 shows how approximations to the Golden Mean were used in eighteenth-century architecture. All sorts of approximations can be found.

The Regular Pentagon

To Construct a Regular Pentagon Given the Length of Side

Draw the base AB and extend it to C (see Fig. 2.53). Draw a semicircle with a radius AB and centre B as shown. Divide the half circumference into five equal parts **by trial**. Take what you estimate to be one fifth in your compass and step it round. Increase or reduce the distance in the compass and step round again until you get it right. Join B3. Then AB3 is one corner of a regular pentagon. Therefore triangle AB3 can be repeated all round to give the other corners. With centre A, draw an arc with a radius of AB. With the centre B and the radius A3, cut this arc to give corner C. With a radius AB make arcs from centres C and 3 to cross at D. Draw 3D and DE to complete the pentagon.

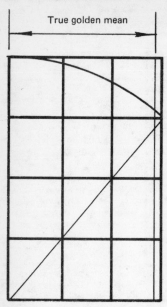

Fig. 2.52. Approximation of the Golden Mean used in eighteenth century glazing bars.

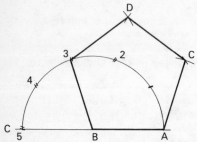

Fig. 2.53. To construct a regular pentagon given the length of side

Other regular polygons can be drawn by this method. Always divide the semicircle into the same number of equal parts as there are sides and draw through the **second** one.

The Regular Hexagon

To Construct a Regular Hexagon Given the Length of Side

Construct the hexagon using a 60° set square on a T square in the order shown (see Fig. 2.54). This is the fastest of all methods.

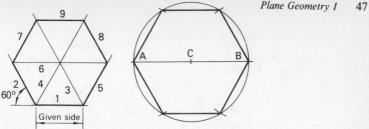

Fig. 2.54. To construct a regular hexagon given the length of side
Fig. 2.55. To construct a regular hexagon inside a circle

To Construct a Regular Hexagon Inside a Circle

Draw the circle and a diameter AB. Set the compass to the length of the radius AC and mark off from the ends of the diameter. (See Fig. 2.55.)

To Construct a Regular Hexagon Given the Distance Across the Flats (*written as A/F*)

Draw the base line (1) and the parallel side (2) the distance A/F away (see Fig. 2.56). Bisect length AF and draw the centre line (3). Draw lines at 60° through any point O on the centre line. Then BC is one side of the hexagon. Complete the hexagon.

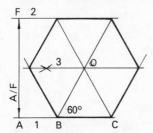

Fig. 2.56. To construct a regular hexagon given the distance across the flats (written as A/F)

The Regular Octagon

To Construct a Regular Octagon Given One Side

Construct in number order (as shown in Fig. 2.57) using a T square and a 45° square.

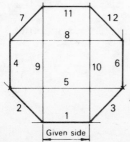

Fig. 2.57. To construct a regular octagon given one side

To Construct a Regular Octagon in a Square (I)

This construction must be **learnt** since nobody is likely to work it out. Remember it as a Maltese Cross.

Method (see Fig. 2.58.) Draw the diagonals. Take a radius equal to half the diagonal and draw arcs from each corner to cut the square at A, B, etc. Join up.

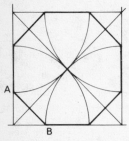

Fig. 2.58. To construct a regular octagon in a square (I)

To Construct a Regular Octagon in a Square (II)

Draw the square. Find the centre by drawing the diagonals. Draw the circle inside the square and draw tangents to the circle at the point where the diagonals cut it. (See Fig. 2.59.)

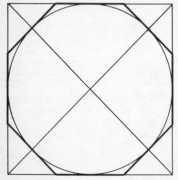

Fig. 2.59. To construct a regular octagon in a square (II)

To Construct a Regular Octagon in a Given Circle

Draw a diameter (1). Bisect and draw diameter (2). Re-bisect angles for 3 and 4. Join up. (As shown in Fig. 2.60.)

Fig. 2.60. To construct a regular octagon in a given circle

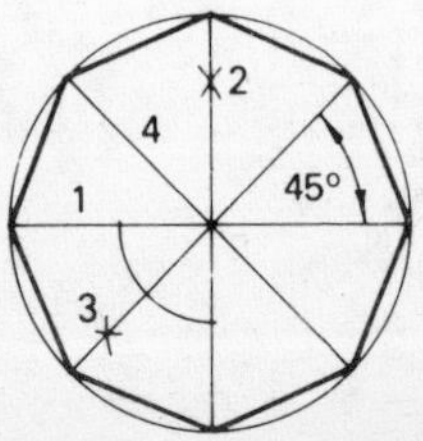

To Construct a Regular Octagon Given the Length of Side

(See Fig. 2.61.) Draw one side AB and an adjacent side BC at 135° (see list in Fig. 2.62). Bisect both sides to cross at O. Draw a circle with centre O to pass through the three corners. Step off the other sides.

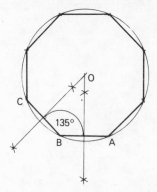

Fig. 2.61. To construct a regular octagon given the length of side

The Angles at the Corners of Regular Polygons

The angles shown in Fig. 2.62 are worth learning as they can help with constructions – as Fig. 2.61 has just demonstrated. Always print in the **size of the angle** when you draw it from memory, or the examiner will not know how you constructed the figure.

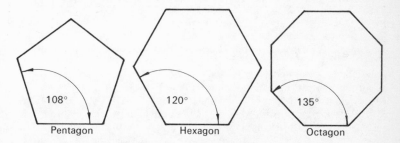

Fig. 2.62. The angles at the corners of regular polygons

Formula for the Angle at the Corner of any Regular Polygon

Take any point P inside the figure and join it to all the corners (see Fig. 2.63). This gives as many triangles as there are sides. The angles in each triangle add to 180° or two right angles.

Therefore the number of degrees in all the triangles equals:

Number of sides × two right angles

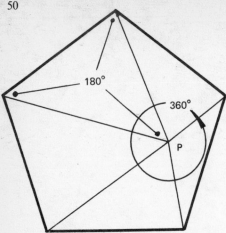

Fig. 2.63. Formula for the angle at the corner of any regular polygon

The total of all the angles meeting at P equals 360° or four right angles. The total of all the angles at the outside corners equals the number of sides × two right angles minus four right angles.

Therefore the angle at the corner of a regular polygon equals:

$$\frac{\text{No. of sides} \times 2 \text{ right angles} - 4 \text{ right angles}}{\text{No. of sides}}$$

Example The angle at the corner of a regular pentagon equals:

$$\frac{5 \times 2 \text{ right angles} - 4 \text{ right angles}}{5} = \frac{(10 - 4) \times 90°}{5} = \frac{6 \times 90°}{5} = \frac{540°}{5} = 108°$$

To Draw a Nest of Regular Polygons, Given the Side

Draw the base AB and bisect it (see Fig. 2.64). Draw a line from A at 45° to cut the bisector at 4. With the centre 4, draw a circle through A and B. This circle will contain a square with sides equal to AB. Draw a line from A at 60° to cut the bisector at 6. With the centre 6, draw a circle through A and B. This circle will contain a regular hexagon with sides equal to AB. Bisect the line 4–6 at 5. This circle (with 5 at the centre) will contain a regular pentagon side AB. Repeat up to centre 9 or 10. After this the method becomes inaccurate.

This construction is of little value in the real world but it is fun to draw.

Questions on Polygons

1 Construct a regular hexagon of 30 mm side using a T square and set square.
2 Construct a regular hexagon in a circle diameter 60 mm.
3 Construct a regular octagon in a 70 mm diameter circle.
4 Construct a regular octagon in a square of 60 mm side.

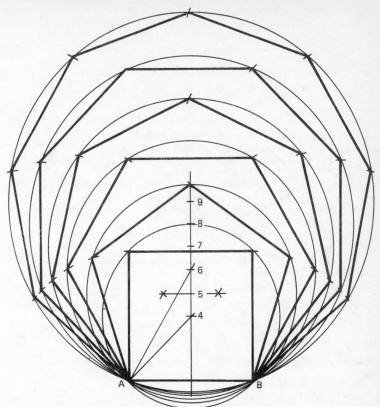

Fig. 2.64. To draw a nest of regular polygons, given the side

5 Construct a regular hexagon which is 50 mm across the flats (A/F).
6 Construct a regular pentagon of 50 mm side by trial.
7 Construct a regular pentagon by setting out the corner with a protractor **and labelling it.**
8 Construct a regular heptagon of 40 mm side by trial.
9 Construct the largest possible regular hexagon inside a 70 mm diameter circle.
10 Construct the largest regular pentagon that will fit in a 70 mm diameter circle.

Circumference of a Circle

Finding the Circumference of a Circle by Rankin's Method

Fig. 2.65 shows a very easy method of finding the circumference of a circle by drawing, and, once learnt, can often be useful.

Method Draw a quadrant and the tangent as shown. Divide the quarter circumference into four equal parts. With centre O and radius 0–1, draw an arc to cut the tangent at 5. With centre 5 and radius 5–4, cut the tangent at 6. Then 0–6 is a quarter circumference.

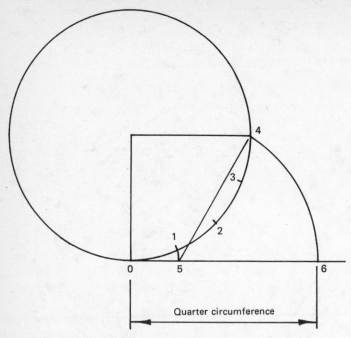

Fig. 2.65. Finding the circumference of a circle by Rankin's Method

A Second Method of Finding the Circumference of a Circle

Draw tangent 3D long. Draw OA at 30° and a horizontal AC. Join CB. Then CB = πD. (See Fig. 2.66.)

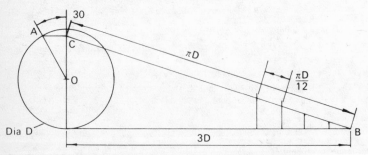

Fig. 2.66. A second method of finding the circumference of a circle

Circle Constructions

To Draw an Arc Radius 60 mm to Pass Through Two Given Points A and B

Draw two arcs radi 60 mm from centres A and B to cross at C. C is the required centre. (See Fig. 2.67.)

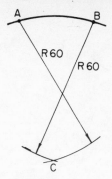

Fig. 2.67. To draw an arc with a radius of 60 mm to pass through two given points A and B

To Draw a Circle Radius 30 mm to touch a Straight Line at Point D

Draw a perpendicular CD to the line AB (see Fig. 2.68). Measure 30 mm along CD from D. This is the centre.

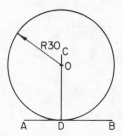

Fig. 2.68. To draw a circle with a radius of 30 mm to touch a straight line

To Draw a Circle Radius 30 mm to Touch a Line AB and a Point C

Draw a line DE parallel to the line AB and 30 mm from it as shown in Fig. 2.69. Draw an arc with a radius of 30 mm from the centre C to cut DE at O. This is the centre.

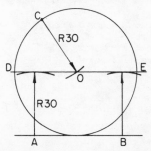

Fig. 2.69. To draw a circle with a radius of 30 mm to touch a line AB and a point C

To Draw a Circle Radius 20 mm to Touch Two Lines AB and CD which do not Meet on the Paper

Draw lines parallel to AB and CD and 20 mm away from them to cross at O. Then O is the centre of the required circle. (See Fig. 2.70.)

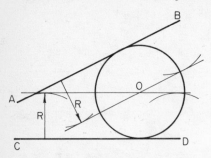

Fig. 2.70. To draw a circle radius 20 mm to touch two lines AB and CD which do not meet on the paper

To Draw a Circle to Pass Through Three Points A, B and C

Bisect the chords AB and BC to cross at 0 (see Fig. 2.71). This is the centre.

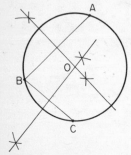

Fig. 2.71. To draw a circle to pass through three points A, B and C

To Draw the Smallest Circle which will Touch and Enclose Two Given Circles with Centres O¹ and O²

Join O^1 O^2 and extend to cut the circles at A and B (see Fig. 2.72). The enclosing circle will pass through A and B. Bisect the line AB at O^3. Draw the circle centre O^3 and radius AO^3.

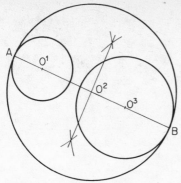

Fig. 2.72. To draw the smallest circle which will touch and enclose two given circles

To Draw an Inscribed Circle in a Triangle

An inscribed circle is one that is scribed (drawn) inside a figure and touches the sides. Bisect any two corner angles of a given triangle to cross at O (see Fig. 2.73). From point O drop a perpendicular to any side at D. OD is the radius.

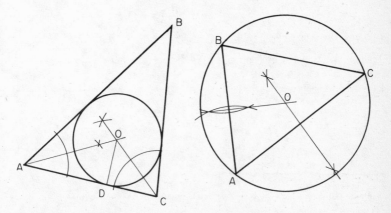

Fig. 2.73. To draw an inscribed circle in a triangle
Fig. 2.74. To draw a circumscribing circle round any triangle

To Draw a Circumscribing Circle round any Triangle

A circumscribing circle passes through all the corners of a triangle. Bisect any two sides of a triangle to cross at O (see Fig. 2.74). Then point O is the centre and OA is the radius.

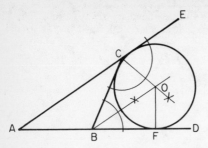

Fig. 2.75. To draw a escribed circle to any triangle

To Draw an Escribed Circle to any Triangle

An escribed circle touches one side of a triangle and the other two sides which have been extended. Let ABC be a triangle with sides AB and AC extended to D and E (see Fig. 2.75). Bisect the angles CBD and BCE to cross at point O. From O drop a perpendicular to either side at F. Then O is the centre of the circle and OF is the radius.

The angles in a circle

The angles subtended by (opposite to) a diameter in a circle are always 90° (see Fig. 2.76).

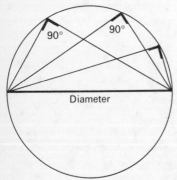

Fig. 2.76. Angles subtended by a diameter in a circle

The angles subtended by a chord in a segment are always equal. They are **smaller than 90°** in the large segment and **larger than 90°** in the small segment. (See Fig. 2.77.)

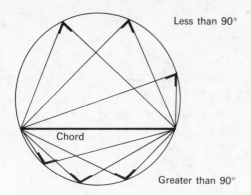

Fig. 2.77. Angles subtended by a chord in a segment

Tangency

A tangent is a line which *touches* a circle (see Fig. 2.78). A tangent is always at 90° to the radius at the point of contact.

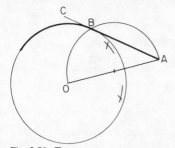

Fig. 2.78. Tangency

To Draw a Tangent to a Circle from a Point A Outside It

(See Fig. 2.79.) Join OA, bisect it and draw the semicircle to cut the circumference at B. Join AB and extend it to C. AC is the tangent.

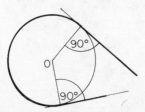

Fig. 2.79. To draw a tangent to a circle from a point A outside it

To Draw a Tangent to Touch Two Circles of Equal Size on the Outside (*externally*)

(See Fig. 2.80.) Join OO¹. Draw perpendiculars at O and O¹ to cut the circles at A and B. Join AB and extend.

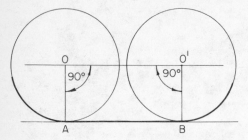

Fig. 2.80. To draw a tangent to touch two circles of equal size on the outside (externally)

To Draw a Tangent to Touch Two Circles of Equal Size (*Internally*)

('Internally' here means that the tangent passes between them.)
(See Fig. 2.81.) Join OO¹ and bisect at A. Bisect OA and AO¹ and draw semicircles as shown. Join CAB which will be a straight line.

Plane geometry continues in Chapter 5 (page 000) but in the meantime we will introduce the topic of solid geometry.

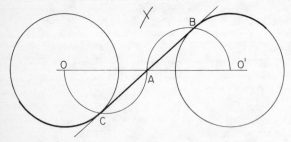

Fig. 2.81. To draw a tangent to touch two circles of equal size (internally)

Plane geometry continues in Chapter 5 (page 77)

3
Solid Geometry I

Visual Aid on Solids

Fig. 3.1 shows picture views of some common three-dimensional solids. These have length, breadth and height. The problem is to show three-dimensional solids on flat, two-dimensional pieces of paper. There are two ways of approaching this problem. **Picture drawings** which show all three dimensions at once and **orthographic projection** which uses several separate views. Each view shows a different face and we put them together in our minds to form a solid. This needs training and is more difficult to understand than a picture but it is easier to add measurements.

Note: 'Ortho' means straight and is used in many words with the senses: straight, rectangular, upright, correct (e.g. orthopaedic – walking correctly (right); orthodontic – teeth correctly placed; orthographic – views projected at right angles).

The Greek 'Perfect' Solids

The Greeks admired things which were regular and obeyed simple, mathematical laws. They admired the cube, the tetrahedron, the octahedron made of equilateral triangles, the dodecahedron with twelve pentagonal sides and the icosohedron with twenty equilateral triangular sides. These they knew as the **Perfect Solids.**

Tetrahedron	4	equilateral triangles
Cube	6	square sides
Octahedron	8	equilateral triangles
Dodecahedron	12	regular pentagons
Icosahedron	20	equilateral triangles

The three-dimensional solids in Fig. 3.1 include 'Perfect' and 'Imperfect' Solids. The perfect cube, tetrahedron and dodecahedron are shown but the other perfect shapes are not. This octahedron is not perfect as the sides are isosceles, and not equilateral, triangles.

Study Fig. 3.1 carefully. You will find these shapes in architecture, in nature (see D'Arcy Thompson's *Nature and Form*), in crystals, in engineering and everywhere in the man-made world.

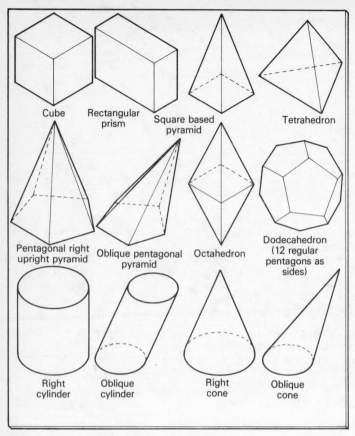

Fig. 3.1. Simple solids drawn in isometric projection: A visual aid

Nine Forms of Projection

Fig. 3.2 shows nine different ways of drawing solid figures. They are shown in detail in later parts of this book.

(*a*) **Isometric projection** Vertical lines stay vertical and horizontal ones slope upwards at 30°.

    Oblique projection The front view is seen as a true shape. Thickness lines slope away at 45°.

(*b*) **Cavalier oblique** Length, breadth and thickness are all drawn full size.
(*c*) **Cabinet oblique** Like Cavalier Oblique but the thickness is halved to improve the appearance.
(*d*) **Single Point Perspective** All lines taper to one point at eye-level.
(*e*) **Two Point Perspective** Vertical lines stay vertical. Horizontal lines taper to two vanishing points on the eye-level.

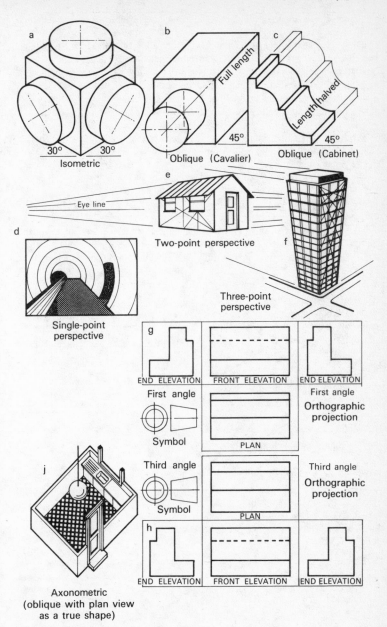

a Isometric

b Oblique (Cavalier)   Full length   30°   30°   45°

c Oblique (Cabinet)   Length halved   45°

Eye line

e Two-point perspective

d Single-point perspective

f Three-point perspective

g
END ELEVATION   FRONT ELEVATION   END ELEVATION
First angle
Symbol
PLAN
First angle
Orthographic projection

Third angle
Symbol
PLAN
Third angle
Orthographic projection

h
END ELEVATION   FRONT ELEVATION   END ELEVATION

j Axonometric
(oblique with plan view
as a true shape)

Fig. 3.2. Nine forms of three-dimensional drawing

(*f*) **Three Point Perspective** Vertical and horizontal lines taper to three vanishing points.

(*g*) **Orthographic projection – First Angle** Views are projected on to surfaces **behind** the object. The surfaces (planes) are then turned flat.

(*h*) **Orthographic projection – Third Angle** Views are drawn on transparent planes **in front of** the object. The planes are then turned flat. This is also called **American projection.**

(*j*) **Axonometric projection** A plan is drawn as a true shape and turned, usually at 45°, 30° or 60°, to the horizontal. Verticals are then projected. Architects like this projection as it gives a true plan and true heights. It can be thought of as an Oblique projection where the *plan* is drawn as a true shape.

Orthographic projection – Third Angle

Fig. 3.3(*a*) shows a picture drawing of a ring made of square material. There is a sheet of glass in front of it and another on top. We look straight at the front and draw the ring on the glass. This is called the **front elevation.** We look down from the top and draw the edge view on the glass. This is called the **Plan**. In Fig. 3.3(*b*) the plan has been hinged upwards. The plan

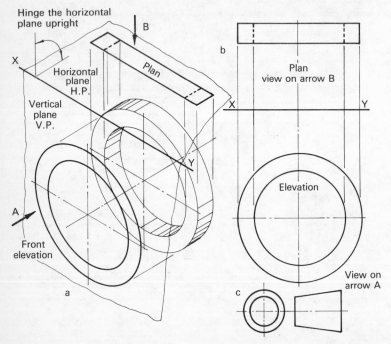

Fig. 3.3. Third Angle projection: A two-view drawing. ((*a*)–(*c*)) (*c*) Symbol for Third Angle projection

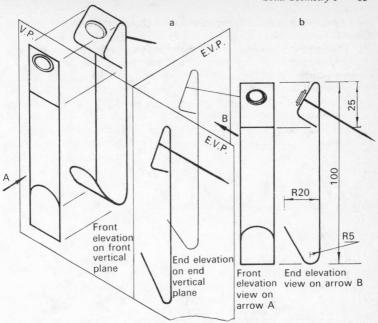

Fig. 3.4. A picture hook in Third Angle projection. ((*a*) and (*b*))

is shown above the elevation because this is how the glass folds. The **symbol** for Third Angle projection is shown in Fig. 3.3(*c*)

In Fig. 3.4(*a*) the plan of a picture hook would not tell us much so a

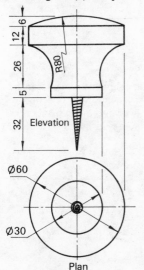

Fig. 3.5. A screw knob
Note: Draw the circles first and project from them

front elevation and end elevation have been chosen. A vertical plane (VP) is in front of the hook and an end vertical plane (EVP) at one side. In Fig. 3.4(*b*) the planes have been folded flat. The XY line is not shown. Some draughtsmen use them and some do not. You may choose which method is more suitable for each drawing.

A screw knob in Third Angle Projection is shown in Fig. 3.5. Work out the position of the knob when it gives these views by holding a similar object and turning it.

Fig. 3.6 shows a casement plate. Again, a third view would not tell us any more.

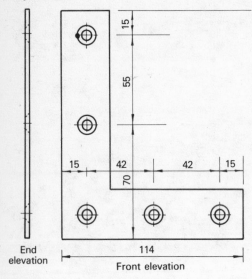

Fig. 3.6. A casement plate in Third Angle projection

Note: All measurements in millimetres

Fig. 3.7(*a*)–(*d*) shows picture drawings of small blocks which can be drawn in two views. Fig. 3.7(*e*) is a freehand orthographic projection of Fig. 3.7(*d*) drawn on squared paper. Setting out drawings on squared paper is very easy. Use it whenever you can for planning and making freehand sketches. It saves time and is clear.

Fig. 3.8 shows a screwdriver drawn in Third Angle projection in two views but a detail of the tip is shown in a third view.

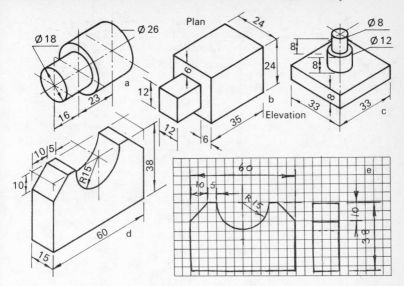

Fig. 3.7. Small blocks which need only two views in orthographic projection. ((a)–(e))

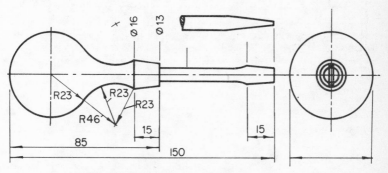

Fig. 3.8. A screwdriver in Third Angle projection

The Unfolding of a Three View Orthographic Projection in Third Angle

Fig. 3.3(a) showed a ring with glass plates in front and above it. The elevation and plan were drawn on the glass. Fig. 3.4(a) showed a picture hook with glass plates at the front and on one side. Fig. 3.9 shows a block with four glass plates. They open out to give three elevations side by side and a plan on top. **In Third Angle, the plan is always on top.**

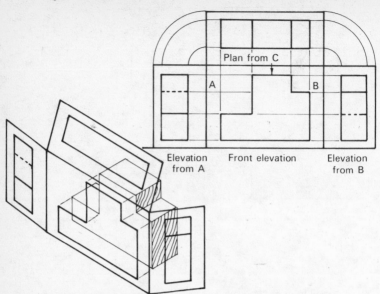

Fig. 3.9. The unfolding of a three-view orthographic projection in Third Angle

Note: Third Angle projection has the advantage of short projection lines

Small drawing blocks (see Fig. 3.10) are very useful individual visual aids. Teachers may like to make up a set of these and give it to the candidates in an end-of-term exam to draw in orthographic projection, isometric projec-

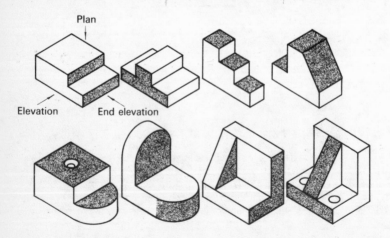

Fig. 3.10. Isometric projections of small drawing blocks

Note: The faces can be painted different colours to make class discussion and the setting of questions easier

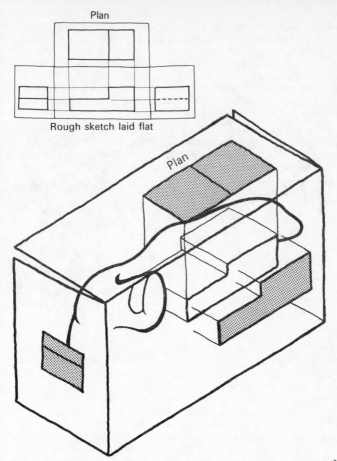

Fig. 3.11. Third Angle projection: Offering up a small block in a torn paper 'room'. The room must be thought of as transparent. A transparent box is ideal.

tion, etc. It will be fresh to everybody. After the exam there will be a class set of drawing blocks for teaching purposes. It is very good for candidates to be able to lay the blocks in different positions and sort out the views. Fig. 3.11 shows a small block being tried in a corner made from torn paper. The blocks repay the time spent on making them by encouraging hours of careful drawing. These blocks have been coloured in one view to help the students to visualise the separate views.

Note: Compare these drawings with the ones in Fig. 3.1 where the inside lines are light and those bordering space are heavy. Which do you prefer?

Questions

1 Make freehand two view orthographic sketches on squared paper of Fig. 3.7(*a*), (*b*), (*c*) and (*d*). Choose the views which will tell you most. Add *all* necessary measurements.

2 Make two view orthographic drawings from your sketches using instruments. You should not refer back to the book. Engineers have to go to a site, make measured sketches, and from them, back in the office, they make finished drawings. They cannot return, half-way across the county, to check a measurement.

Note: Your drawings must always show measurements in **the correct style.** A wrongly placed measurement can cause disaster. Remember that the measurement must be at right angles to the arrow shaft, above it when read from the bottom right-hand corner of the page (see Fig. 1.13). The history of engineering is littered with expensive mistakes caused by careless draughtsmen.

3 Draw Fig. 3.3(*a*) as an elevation and end elevation with the outside diameter 60 mm and then material 10 mm square.

4 Draw Fig. 3.4(*a*) as an elevation and plan with the picture hook lying flat. Estimate the measurements not given and all the curves.

5 Draw the screw knob as shown in Fig. 3.5.

6 Draw Fig. 3.6 with the end elevation on the left.

7 Draw Fig. 3.8 as shown. You will need to look up Tangency and/or your teacher will help you.

8 Draw the blocks shown in Fig. 3.10 so that the shaded sides are the front elevations.

Discussion of solid geometry continues in Chapter 6 (page 102)

4
Isometric Drawing – Pictorial Drawing I

History

Isometric drawing is a form of 'picture drawing' invented in 1820 by Sir William Farish for giving instructions to his laboratory assistants. He called it 'Isometrical Perspective'. 'iso' means equal and 'metric' has to do with

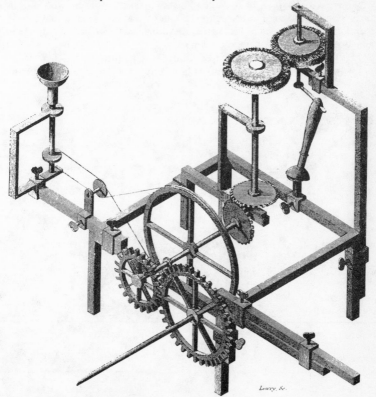

Fig. 4.1. From *Isometrical Perspective*, by Sir William Farish, 1820

measurement. In isometric projection, the height, width and thickness measurements are drawn to their original sizes. All vertical lines stay vertical, while the width and thickness measurements are inclined at 30° to the horizontal.

Farish decided to give a series of lectures on the efficiency of machines and machine parts. He was a professor of Chemistry at Cambridge and nobody had dealt with mechanical efficiency in this way before. There had to be lectures on particular machines but he wanted to examine the function of particular gear wheels, pulley systems, etc. Each lecture required working models and they changed each week.

To save space, Farish had a series of engineering parts made to a small scale. These could be fitted together on bars and frameworks, to make different machines. It was a sort of early Meccano. Each machine was drawn in isometric projection so that his assistants could assemble it (see Fig. 4.1). Set squares had not been invented. However, he developed a slotted rule that did fairly well, and also a series of ellipse templates for drawing circular parts.

Since then, several manufacturing firms have used isometric drawings to explain assembly work to unskilled and semi-skilled workers who do not understand orthographic projection. Similarly, catalogues and instruction manuals, and many advertisements, are in isometric projection since they are so easily understood by the layman.

The Japanese and Persians had been drawing pictures from a high viewpoint for hundreds of years but without any set rules. 'Children on the

Fig. 4.2. 'Children on the Rampage in a classroom while the Teacher looks on impassively'. From *Ehon mote – asoki*, by Slimokobe Shusui. First published 1780. Copyright the British Library.

Rampage' has lines at about 30° so apparently it is in isometric projection (see Fig. 4.2). The book is in a flip-flap style, a long sheet folded like a concertina. Looking through it one sees that the drawings are sometimes at 45°, sometimes at other angles. There is no rule.

The Rules of Normal Isometric Drawing

In normal isometric drawing vertical lines are drawn as verticals but are really sloping towards you out of the paper. Horizontal lines are drawn at 30° to the horizontal. All horizontal lines and all vertical lines are drawn true to length.

Isometric scale, where some sides are shortened, is dealt with later in Fig. 7.34. **It is not normally used.** I have not seen an 'O' level question which includes isometric scale for some time, but it may be asked at 'A' level.

Fig. 4.3 shows a square which is to be drawn in isometric projection.

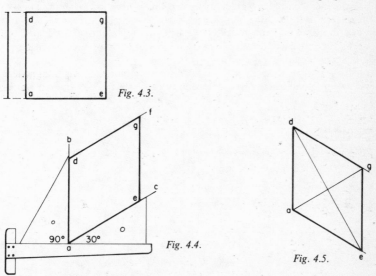

Fig. 4.3.

Fig. 4.4.

Fig. 4.5.

Using a 30°/60° set square on a T square, draw a vertical line ab longer than is necessary (see Fig. 4.4). Draw ac at 30° to the horizontal. Measure ad (the height of the square) along ab and draw df at 30°. Measure ae along ac and draw eg another vertical. Line in adge. Fig. 4.5 shows adge drawn in reverse. Diagonals ag and de in the square are equal but they do not stay equal when drawn isometrically.

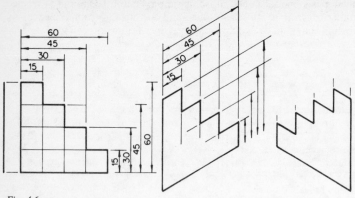

Fig. 4.6.

Repeat the method for Fig. 4.6 but make a network before lining in. Try not to do it in little bits.

Fig. 4.7 shows a cube in orthographic projection (Third Angle).

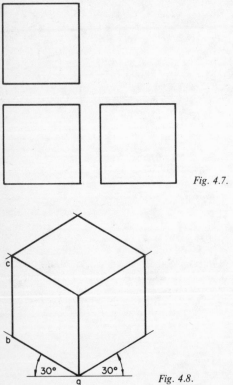

Fig. 4.7.

Fig. 4.8.

Method Draw one face abcd in isometric projection (see Fig. 4.8). Project the other edges at 30°, measure along and join up to form the cube.

Fig. 4.9 shows an orthographic projection of a stepped block.

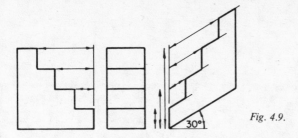

Fig. 4.9.

Method Draw one end in isometric projection. Project the other edges at 30°, measure along and join up (see Fig. 4.10).

Fig 4.11 shows the front and end elevations of a ridged block.

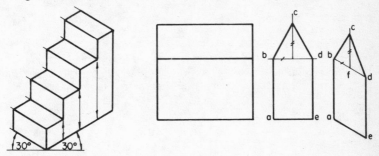

Fig. 4.10.   Fig. 4.11.

Method (See Fig. 4.12.) Draw ae and the verticals ab and ed in isometric projection. Join bd. Bisect bd at f and draw the vertical cf. Join bc and cd. Complete the block (see Fig. 4.13).

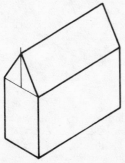

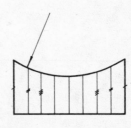

Fig. 4.12.   Fig. 4.13.

Fig. 4.14.

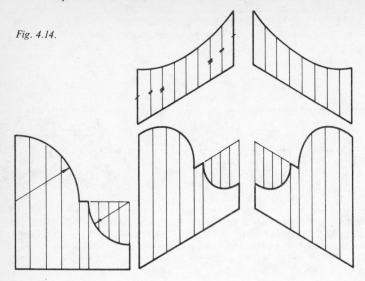

Curves and irregular shapes are drawn by erecting verticals at intervals and then measuring the correct height on each (see Fig. 4.14). Repeat the method, measuring up or down as you choose (see Fig. 4.15).

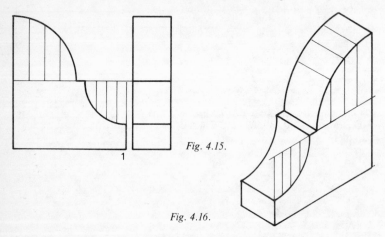

Fig. 4.15.

Fig. 4.16.

Fig. 4.16 shows the orthographic projection (Third Angle) of a wooden moulding. Draw the front elevation in isometric projection (see Fig. 4.17). Project thickness lines at 30°. Measure the thickness all along with a compass and join up. This is far quicker than repeating the first construction as the compass stays the same size.

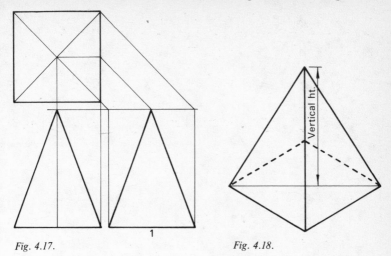

Fig. 4.17.   *Fig. 4.18.*

Fig 4.18 shows a plan and two elevations (Third Angle) of a right (up-right) pyramid with a square base.

Fig 4.19 shows a **square based pyramid**. Draw the square base in isometric projection and its diagonals. Draw a vertical line from the centre of the base. Measure the height of the pyramid **from the centre of the base.** Join up. It is like putting up a tent with one central pole.

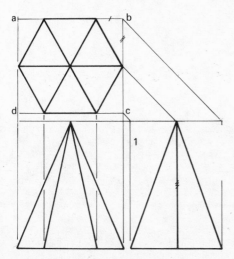

Fig. 4.19. A square based pyramid

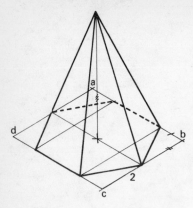

Fig. 4.20. A right hexagonal pyramid

Fig. 4.20 shows **a right hexagonal pyramid.** Draw the rectangle abcd surrounding the hexagon. It is difficult to draw the hexagon direct in isometric projection so we draw the rectangle in isometric first and then measure in. Complete the hexagon. Find the centre. Draw the vertical height from the centre of the base and join up.

Discussion of pictorial drawing (isometric drawing) continues in Chapter 7 (page 123).

5
Plane Geometry II

Continued from Chapter 2 (page 58)

Tangency

To Draw a Tangent to Two Unequal Circles Externally

This cannot be done directly. It has to be tackled in two stages, the first of which is 'Draw a tangent to a circle from a **point** outside it' (see Fig. 2.79).

Method (See Fig. 5.1.) Take R^1 from R^2. Draw a circle centre O^1 with radius $R^1–R^2$. Now we have the construction shown in Fig. 2.79. Join OO^1 and bisect. Draw a semicircle to cut the third circle at A. Then OA is a tangent. Join O^1A and extend to cut circle R^2 at B. Draw OC parallel to O^1B. Join BC. This is the required tangent. Check that ABCO is a rectangle. If it is not, there is a mistake in the construction.

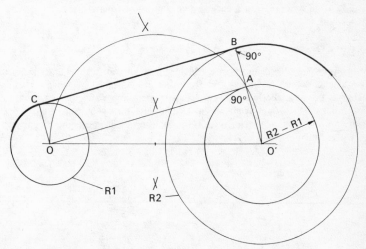

Fig. 5.1. To draw a tangent to touch two unequal circles externally

To Draw a Tangent to Touch Two Unequal Circles Internally

Again it must be turned into Fig. 2.79 first.

Method (See Fig. 5.2.) *Add* R^1 to R^2. Draw a circle with this radius. Join OO^1 and erect a semicircle on it to cut the new circle at A. Join AO^1 to cut circle R^2 at B. Draw AO which is the tangent to circle $R^2 + R^1$. Draw OC parallel to AO^1. Join CB and extend. This is the required tangent. Check that OABC is a rectangle and not a parallelogram. If OABC is a parallelogram your construction is wrong. Check it.

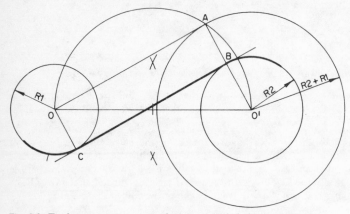

Fig. 5.2. To draw a tangent to touch two unequal circles internally

Tangency Questions

Draw Fig. 5.3(*a*)–(*e*) double size or larger. Set out centre lines and circles. Then decide which tangency constructions are required and draw them **leaving in the constructions.** Line in smoothly, taking particular care with the joins. Fig. 5.3 (*f*) shows a **single-feed supply.** This is a control gate for supplying one article or piece of material at a time to a machine. It works like the escapement of a clock.

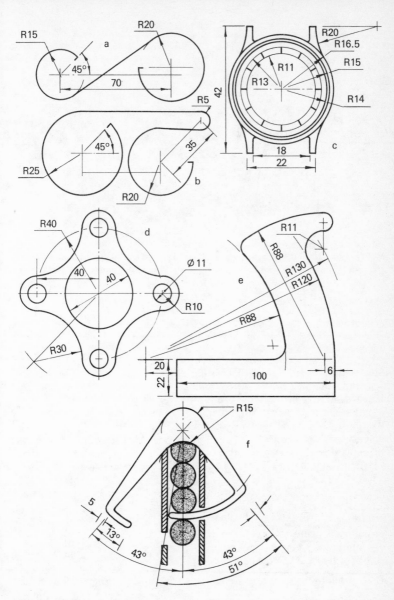

Fig. 5.3. Tangency questions

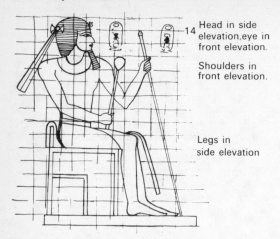

Fig. 5.4. Tuthmosis III 1490–1439 BC
A drawing on a wooden board covered in fine plaster. The rough square grid shows 14 units to the hairline. The grid is in finger widths. It could later be sealed up to any size. (British Museum – from a Theban Tomb.)

Enlargement and Reduction of Plane Figures

Enlargement by Squaring up

This is the oldest method of enlarging plane figures and still the most common. Ancient Egyptian artists had to draw according to strict rules. A man was always in the same proportions. Each part of the body was a certain number of 'fingers' or 'palms' or 'cubits' long.

An artist could make his own ruler by marking from his own body. The rulers varied in size but this did not matter. The drawings an ancient Egyptian draughtsman had to make were in proportions, not absolute sizes. A small sketch drawn on squares a finger wide, could be copied on squares a palm or a cubit wide and the proportions kept the same (see Fig. 5.4).

The Egyptian Scale of Measurement

Four fingers made a palm.
Six palms made a cubit.
Seven palms made a Royal cubit.

These proportions were very useful of course. There were no absolute sizes.

Robert Fludd illustrates reducing by using a square net like a tennis racket (see Fig. 5.5). These contraptions were used widely by artists for drawing landscapes, figures in unusual positions (descending from heaven for instance) and for complicated buildings. Canaletto used this system: each square in turn was given equal attention, so that his pictures have an even, bland accuracy.

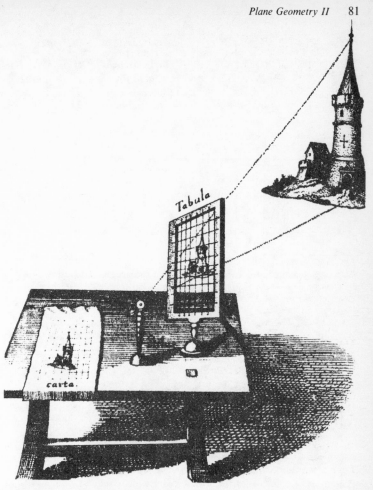

Fig. 5.5. Reducing by looking through a spy hole through a string grid and drawing on squared paper. Robert Fludd 1574–1637.

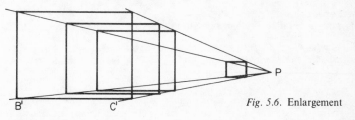

Fig. 5.6. Enlargement

Enlargement

A cinema projector enlarges a frame of a film by radiating it from a point (the projection lamp). Move the screen further from the projector and the picture becomes larger (see Fig. 5.6).

To Enlarge a Rectangle ABCD so that the Sides Double in Length

(See Fig. 5.7). Join PD and extend. Mark D^1 so that $PD = DD^1$. Draw $A^1B^1C^1D^1$ parallel to ABCD. PA^1 is double PA so A^1D^1 is double AD. All the sides of $A^1B^1C^1D^1$ have been doubled in length. Lines may also be projected from a point P inside the figure as shown in Fig. 5.8.

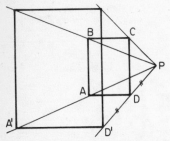

Fig. 5.7. To enlarge a rectangle ABCD so that the sides double in length

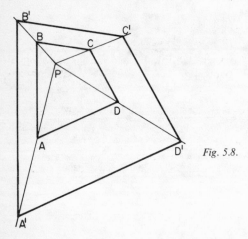

Fig. 5.8.

Enlarging by Similar Triangles or Similitude

Similar triangles have the same angles and their sides are in the same proportions. In Fig. 5.8 for example, PAD and $P^1A^1D^1$ are similar triangles.

To Enlarge a Polygon ABCDE so that the Sides are Increased in the Proportion 3:2.

(See Fig. 5.9.) Extend AB to F. Draw a straight line from A below AB. Mark off three equal spaces. Join 2B. Draw from 3 parallel to 2B to cut AF at B^1. Then AB^1 and AB are in the proportion 3:2. Draw through AC and extend. Draw BC^1 parallel to BC. Then triangle ABC is similar to

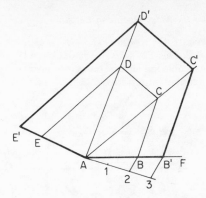

Fig. 5.9. To enlarge a polygon ABCDE so that the sides are increased in the proportion 3:2

triangle AB^1C^1 and all the sides are in the required proportions. Draw C^1D^1 parallel to CD and D^1E^1 parallel to DE.

Reduction in Proportion

To Reduce a Polygon ABCDEF in the Proportion 4:5

(See Fig. 5.10.) Draw any straight line from A and mark off five equal divisions. Join 5B. Draw $4B^1$ parallel to 5B. AB is divided in the proportion 4:5 at B^1. Join AC. Draw B^1C^1 parallel to BC. Join AD and continue round the polygon.

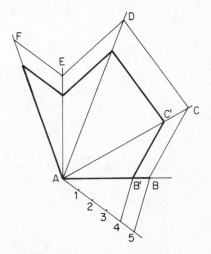

Fig. 5.10. To reduce a polygon ABCDEF in the proportion 4:5

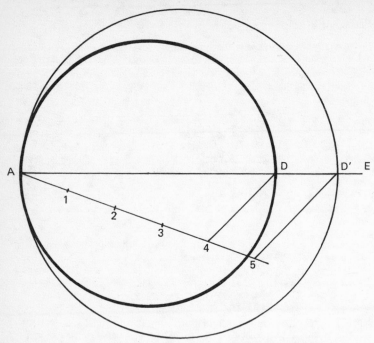

Fig. 5.11. To draw a circle with a diameter in the proportion 5:4 to another circle

To Draw a Circle with a Diameter in the Proportion 5:4 to Another Circle
Let a circle diameter AD be the smaller circle. Enlarge as shown in Fig. 5.11 to AD[1]. This is the enlarged diameter.

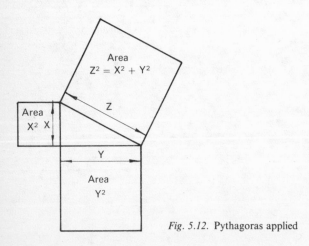

Fig. 5.12. Pythagoras applied

Pythagoras Applied

In any right angled triangle, the sum of the areas of the squares on the two short sides is equal to the area of the square on the hypotenuse (the long side). This theorem by Pythagoras (see Fig. 5.12) is still one of the most important in all plane geometry. You will find Pythagoras hiding in dozens of questions.

The Same Rule Applies to Circles and Semcircles (See Fig. 5.13)

The area of the circle on diameter x
plus   = The area of the circle on diameter z
The area of the circle on diameter y

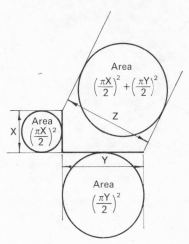

Fig. 5.13. The same rule applies to circles and semicircles

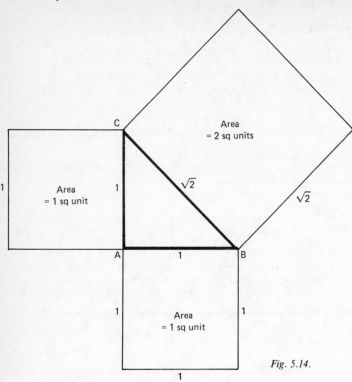

Fig. 5.14.

The right angled isosceles triangle ABC has two sides equal to 1 and the hypotenuse equal to the square root of 2 (see Fig. 5.14). This is a special case of Pythagoras and we can use this fact to double and halve areas.

To Double the Area of a Triangle ABC

Draw a right angled isosceles triangle ABD with AB equal to BD (see Fig. 5.15). Extend AB. With centre A, swing radius AD round to F. Draw FG parallel to BC. Area AFG equals twice area ABC.

Proof: ABC and AFG are similar triangles. Let AB equal x. Imagine a square drawn on side AB. Its area will be x^2. Imagine a square drawn on AF. Its area will be $2x^2$ because AF equals the square root of $2x^2$.

The areas of similar triangles ABC and AFG must be in the same proportion. Therefore area AFG must be double area ABC. Circle cente O^2 has twice the area of circle centre O^1 because diameters PQ and PR are in the

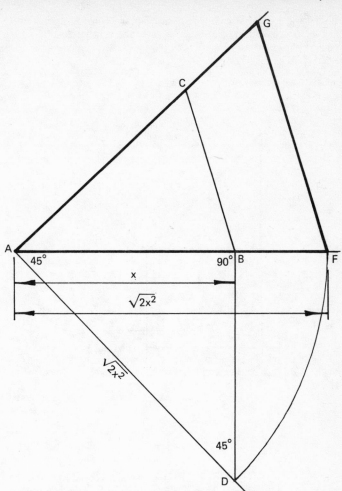

Fig. 5.15. To double the area of a triangle ABC

proportion x and $\sqrt{2x^2}$. This is true of all similar figures in similar positions. Fig. 5.16 shows that the same construction applies to all similar figures.

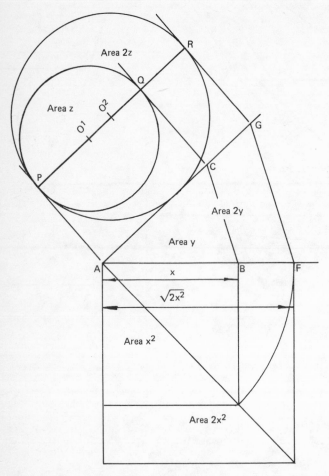

Fig. 5.16.

To Halve the Area of a Polygon ABCDEF

(See Fig. 5.17.) Radiate lines from A through all the other corners. Draw a right angled triangle ABG so that AB is the hypotenuse. With centre A swing AG to B^1 on AB. Draw B^1C^1 parallel to BC. Area AB^1C^1 is half the area of ABC. Project round similarly to D^1, E^1, F^1. Then $A^1B^1C^1D^1E^1F^1$ will be half area ABCDEF.

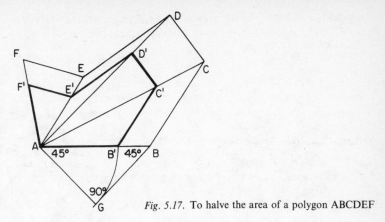

Fig. 5.17. To halve the area of a polygon ABCDEF

To Draw Square Roots

Fig. 5.18 shows triangle ABC with AB = 1 metre and AC = 2 metres. Angle BAC = 90°. The area of square ABDE = 1 square metre and the

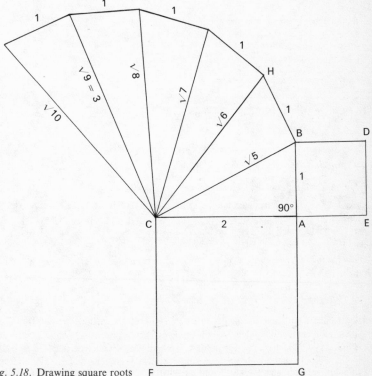

Fig. 5.18. Drawing square roots

area of ACFG = 2 × 2 = 4 square metres. By Pythagoras, the square on BC must equal 4 + 1 square metres = 5 square metres. Therefore BC must have a length of $\sqrt{5}$ m.

In triangle CBH, the square on CH equals the sum of the squares on CB and BH or square metres. Therefore CH = $\sqrt{6}$m.

This process can be repeated as shown. Examination questions asking for unusual square roots are not common today but they turn up occasionally.

The Simplification of Areas

Areas are easier to calculate if they are squares or rectangles so we often change complicated shapes into **simpler ones of the same area.**

To Simplify the Area of Parallelogram ABCD

(See Fig. 5.19.) Draw verticals from A and B. Extend CD to F. Area ADF equals area BCE. Therefore rectangle ABEF equals rectangle ABCD in area. (Rectangle ABED is common to each.) The shaded triangle BCE has slipped across to ADF. The area of a parallelogram equals base × vertical height.

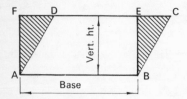

Fig. 5.19. To simplify the area of parallelogram ABCD

To Simplify the Area of Triangle ABC

Area of a triangle = $\frac{1}{2}$ base × vertical height.

(See Fig. 5.20.) Draw vertical height and bisect it at D. Draw perpendiculars from A and B. Draw FDE parallel to the base AB. Triangle DGC = BEG and DCH = HFA. Therefore area ABEF = area ABC.

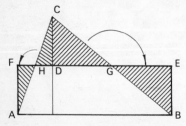

Fig. 5.20. To simplify the area of triangle ABC

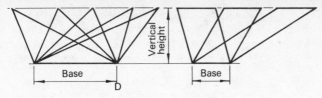

Fig. 5.21.   *Fig. 5.22.*

All the triangles in Fig. 5.21 have the same area as they have the same base and the same vertical height.

All the parallelograms in Fig. 5.22 have the same area as they have the same base and the same vertical height.

To Reduce the Quadrilateral ABCD to a Triangle of the Same Area.

(See Fig. 5.23.) Join BC. We have cut the quadrilateral into two triangles. Extend the base AB to F. Draw CG parallel to DB. Join DG. Now think of DB as a base and CG parallel to it. Triangles DCB and BGD have the same base DB and the same vertical height DJ, and **must be the same area.** ABCD has the same area as triangle AGD.

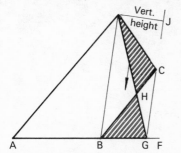

Fig. 5.23. To reduce the quadrilateral ABCD to a triangle of the same area

Another Example of Reduction

(See Fig. 5.24.) Triangle GFD has the same area as the polygon ABCDE. Again, the base has been extended and the top corners have slid down.

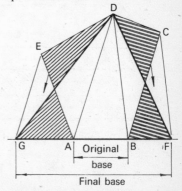

Fig. 5.24. Another example of reduction

Figures with Pieces Cut out of Them

When figure ABCD has a triangle AED cut out of it to give the figure ABCDEA (see Fig. 5.25), we can treat the missing triangle in the same way as above.

Join DA. Draw EF parallel to DA. Join DF. Then triangle AOF is equal in area to ODE. AOF has slipped up into ODE. Thus FDCB has the same area as ABCDE.

Remember that when an **inlet** has been cut into a figure, the base becomes **shorter**. When a **headland** sticks out of a figure, the base becomes **longer** (see Fig. 5.26). Area of ABCDE = Area of FCG.

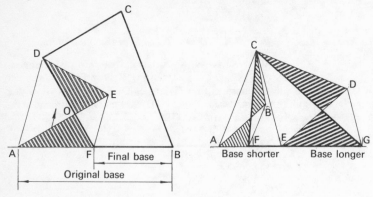

Fig. 5.25. Figures with pieces cut out of them   *Fig. 5.26.*

To Convert a Rectangle ABCD into a Square

(See Fig. 5.27.) Extend the base and swing the vertical height BC on to it at E. Bisect AE and draw the semcircle. Extend BC to F. Then BF is the **mean** of the rectangle. Draw the square BFGH. The area of BFGH equals the area of ABCD.

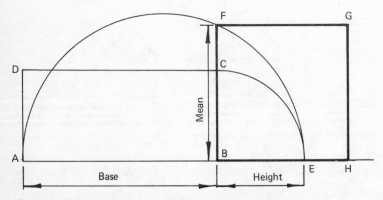

Fig. 5.27. To convert a rectangle ABCD into a square

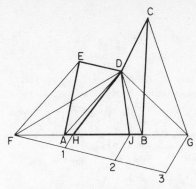

Fig. 5.28. Reduce polygon ABCDE to square of equal area

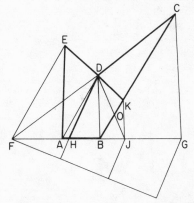

Fig. 5.29. Reduce ABCDE to triangle FCG

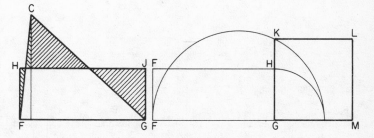

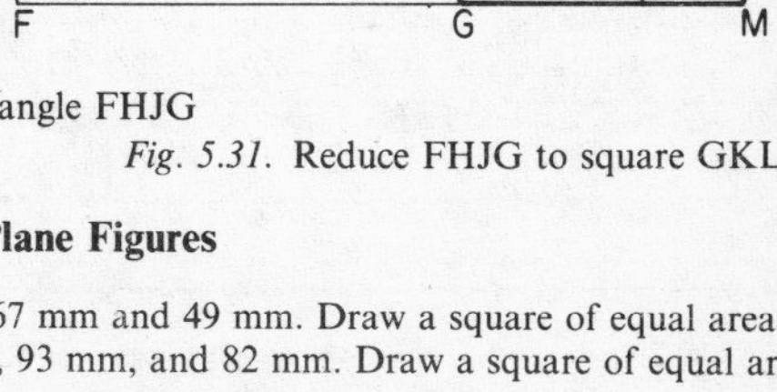

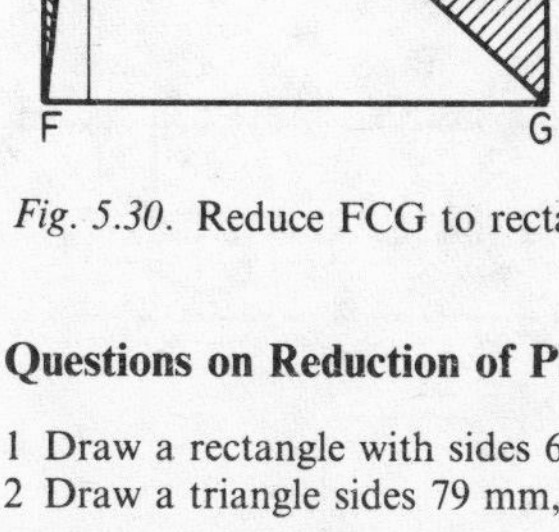

Fig. 5.30. Reduce FCG to rectangle FHJG

Fig. 5.31. Reduce FHJG to square GKLM

Questions on Reduction of Plane Figures

1 Draw a rectangle with sides 67 mm and 49 mm. Draw a square of equal area.
2 Draw a triangle sides 79 mm, 93 mm, and 82 mm. Draw a square of equal area.

3 Draw a regular pentagon of 38 mm side and reduce it to a square of equal area.
4 Draw Fig. 5.23 with AB = 55 mm, BC = 54 mm, DC = 41 mm, DA = 100 mm and BD = 75 mm. Reduce to a square of equal area.
5 Draw Fig. 5.24 with AB = 34 mm, CB = 68 mm, CD = 32 mm, DE = 51 mm, EA = 55 mm, AD = 79 mm and DB = 76 mm.

Any of the other drawings in this section can be drawn in the same manner at double size or larger and reduced to simpler figures.

Question 6 is more difficult but is based on the same type of geometry. It is 'O' level standard, on Grade 1 GCSE.

Question ABCDE is a field. Divide it into three equal areas.

Solution (a) Fig. 5.32(*a*). Change the shape into a triangle FDG of equal area. Divide GF into three equal parts at H and J. Join DH and DJ. Then DEAH = DHJ = DJBC in area.

Solution (b) In Fig. 5.32(*b*) the shape ABCDE is different. GF is divided at H and J. H is in the field so AHDE is one third of the field. J is not part of the field so DHJ is not the one third we need but is the right area. HBD is part of the field. We must add on an area equal to BJD. Draw JK parallel to DB. Then BKD equals BJD. Then HBKD equals HJD and is one third. DKC is the final third.

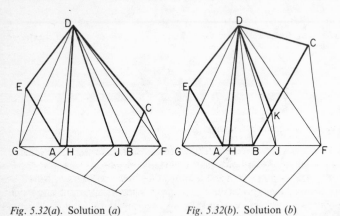

Fig. 5.32(a). Solution (*a*)   *Fig. 5.32(b).* Solution (*b*)

Drawing by Coordinates

We sometimes use coordinates to set out irregular shapes. The profile of a line of cliffs can be measured and set out like a graph. Table 5.1 shows distances from the left (the east in this case) to the right and each height.

Set out the sea level and draw the shape of the cliff. Scale 1 millimetre : 1 metre (1 : 1,000).

Table 5.1 Profile of the cliffs in figures

Height above sea level in metres																			
10	17	11	18	23	32	41	37	33	39	35	32	31	25	20	17	13	9	6
0	10	20	30	40	50	60	70	80	90	100	110	120	130	140	150	160	170	180
East							Distance in metres											West

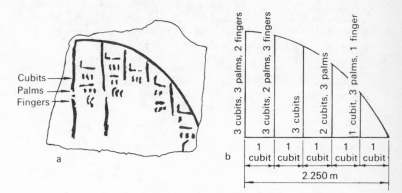

Fig. 5.33(a). An Egyptian architect's drawing *Fig. 5.33(b).* A modern translation

An Egyptian Architect's Drawing

Fig. 5.33(a) is a copy of one of the earliest drawings using coordinates. It was made to instruct the builders of a barrel vault of sun-dried bricks for a tomb roof. The sketch was made on a flake of limestone with a brush. The builders would have known that the uprights represented cubit intervals. The heights were written in cubits, palms and fingers. A cubit is the length from the elbow to the tip of the middle finger. There were, at this time, six palms to the cubit (later there were seven) and four fingers to a palm. It is a convenient system of measuring, as the foreman always carried his 'ruler' about with him. Fig. 5.33(b) shows the same measurements worked out in metric. The sizes would really have varied according to the length of the foreman's arm, but the shape would have been the same regardless.

The Cross section of a River and Jetty

The river (as shown in Fig. 5.34) needs to be dredged so that the ship can reach the jetty fully laden, at all states of the tide, with two metres of water below the keel. Draw the cross section of the river and show the line to which it must be dredged.

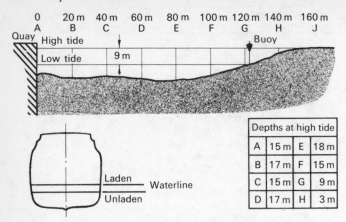

Depths at high tide			
A	15 m	E	18 m
B	17 m	F	15 m
C	15 m	G	9 m
D	17 m	H	3 m

Fig. 5.34. The cross section of a river and quay

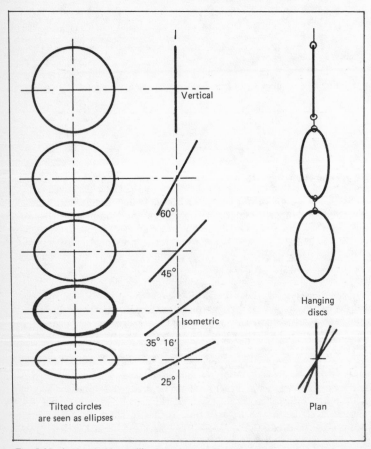

Fig. 5.35. A visual aid on ellipses

Ellipses

A visual aid on ellipses

A tilted circle is seen as an ellipse (see Fig. 5.35). The diameter becomes the **major axis.** The centre line at right angles becomes the **minor axis.** The minor axis varies from the length of the diameter to zero. An ellipse can vary from a circle to a straight line. The hanging disks are useful for demonstrating the characteristics of ellipses. They hang vertically but can be drawn out horizontally.

Ellipse Teaching Aid

This consists of a circle in hardboard which pivots along the diameter AA on a wire frame (see Fig. 5.36). The frame and circle can be hung up as a general visual aid and be hand held for demonstrations.

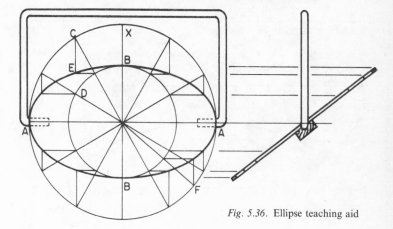

Fig. 5.36. Ellipse teaching aid

The Concentric Circle Method of drawing an ellipse

Method (see Fig. 5.36). Draw the major axis AA and the minor axis BB. Draw concentric circles through AA and BB. Divide the circles into twelve equal parts using a 30°/60° set square. Points C and D are **similar** points on the two circles. Draw vertically down from C. Draw horizontally from D to E. This is a point on the ellipse. Repeat for other points and join freehand. The visual aid illustrates the movement of the different points. We have drawn lines at 30°, 60° but any radial line can be used if preferred (see F). However, 30°, 60° are quick and easy.

Joining Points Freehand

This needs a lot of practice. Make a series of dots and practise joining

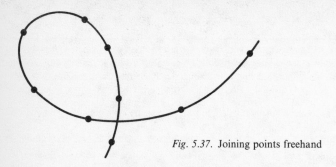

Fig. 5.37. Joining points freehand

them freehand (see Fig. 5.37). Work lightly and feel your way to the line. Do not make a solid, deep line that is difficult to rub out. Be very careful if you use french curves or flexible curves. Most people draw smooth curves with them **which miss all the points.**

The Spider's Web Method: Visual Aid

Draw a rectangle enclosing the major and minor axes (see Fig. 5.38). Divide AB (equal to half the height of the minor axis) into four equal parts and

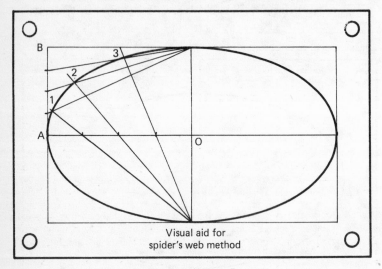

Visual aid for
spider's web method

Fig. 5.38. The spider's web method: Visual aid

draw radiating lines from the top of the minor axis. Divide AO (half the major axis) into the **same number** of equal parts. Draw radiating lines from the bottom of the minor axis to cut the first radiating lines at 1, 2 and 3. Join up freehand. Repeat the procedure for the other quarters of the ellipse. Remember – **radiate from the top and bottom of the minor axis.**

The Trammel Method

This is a useful practical method. Take the straight edge of a piece of paper or card. Mark from one end E the lengths of half the major axis (M) and half the minor (m). **They must both be marked from the same end** (see Fig. 5.39–5A).

Method 1 Place the card so that M is on the minor axis and m is on the major axis. Mark the position of the end E. This is a point on the ellipse. Repeat and join the points.

Method 2 Mark out a trammel as at 5B, with the half major and half minor axes **end to end**. Hold the ends T and S on the minor and major axes extended and mark point M at 4 and 5 etc.

Visual Aid to Show Trammel Method

A board is marked out with an ellipse and its axes (see Fig. 5.39). Two slots run along the axes and a trammel rod (with pegs at M and m) slides

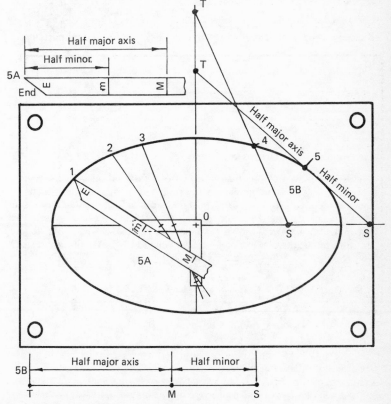

Fig. 5.39. Visual aid for the trammel method

along the slots. End E traces out the ellipse. For Method 2 (see above) another rod is hand held.

The Focal Point Method

(See Fig. 5.40.) Focal points f^1 and f^2 are two points on the major axis equidistant from the ends. The distance from f^1 to any point on the ellipse and on to f^2, is always equal to the length of the major axis. This is true even of point A because Af^1 is covered twice and $Af^1 = f^2B$.

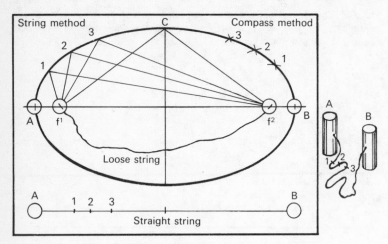

Fig. 5.40. Visual aid for the focal point method

The Geometrical Construction

(See the upper part of Fig. 5.40.) Draw the major and minor axes. With a centre at the top of the minor axis C and a radius equal to half the major axis, cut the major axis at f^1 and f^2. These are the focal points. Divide the half major axis as shown (removed below) at 1, 2, 3. Take the radius A1 and make an arc from centre f^2. Take the radius 1B and make an arc centre f^1 to cut the first arc at 1. Repeat for radii A2, A3. Join points B,1 2, 3, C. Draw the other quarters similarly.

A Practical Method: To Set Out an Oval Flower Bed in a Lawn

(See the lower part of Fig. 5.40.) This is the way a gardener or a man setting out a piece of ornamental paving, might tackle the problem of drawing an ellipse. It works very well with a piece of cotton and some dress pins. I have seen it used at 'A' level by a student who forgot every other method. And he passed!

Method Set out the major axes with lengths of string and pegs. Take another piece of string the length of the major axis and tie a peg at each end.

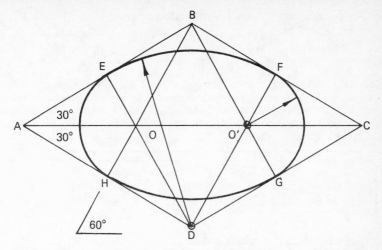

Fig. 5.41. An approximate method of drawing an ellipse inside an 'isometric square'

(In the Visual Aid Fig. 5.40 this is shown as AB loose at one side and stretched tight below.) Fold the string in half and mark out f^1 and f^2 by drawing an arc from the top of the minor axis. Push in the pegs at f^1 and f^2. The string will hang loose as shown. Stretch the string taut as at 1, 2 and 3 and mark out the ellipse. Coloured tags can be added to the string on the Visual Aid as shown.

An Approximate Method

This method is convenient and quick but it is only approximate and should never be used for drawing concentric ellipses as this will exaggerate the faults.

Method Draw a diamond shape with a 30°/60° set square. In Chapter 7 we shall call this an Isometric Square. This is labelled ABCD. From B and D draw lines with a 60° square to cross AC at O and O^1. E, F, G, H will all be the centres of the sides. With centre O and radius OE, draw the arc to H. With centre D and radius DE draw arc EF. Complete the ellipse similarly.

6
Solid Geometry II

Continued from Chapter 3 (page 68)

First Angle and Third Angle Projections

Fig. 6.1 shows a vertical plane and a horizontal plane crossing each other to give four quadrants. The object is in the Third Quadrant. The plan is drawn on the horizontal plane and swings up **above** the elevation. (See Fig. 3.3.)

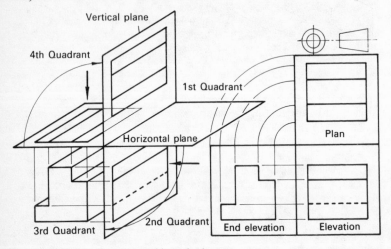

Fig. 6.1. Third Angle orthographic projection

Fig. 6.2 illustrates First Angle projection with the object in the first quadrant. The plan swings **below** the elevation. It is very important to understand the difference between First Angle and Third Angle (or American) projection.

Questions

Answer all questions in First Angle projection.

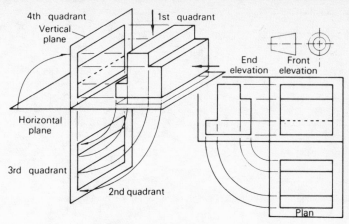

Fig. 6.2. First Angle orthographic projection

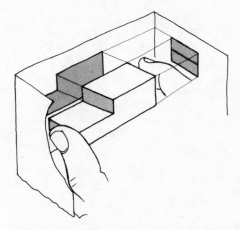

Fig. 6.3. Offering up a small block in a torn paper 'room'

1 Copy Fig. 6.4 double size, scaling from the book.
2 Draw three views in First Angle projection of some of the blocks in Fig. 3.10 with the shaded surfaces as end elevations. Draw them freehand on squared paper.
3 Draw six views of one block using instruments.
4 Draw Fig. 6.6 using your own sizes. Add four important measurements.
5 Draw three views of Fig. 6.7.
6 Draw four views of Fig. 6.8. Move the measurements well away from the drawing.
7 Draw three views of Fig. 6.9 with the plate lying horizontally.

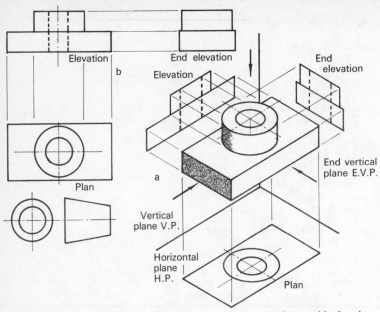

Fig. 6.4. First Angle Projection The views have been opened out with the plan below

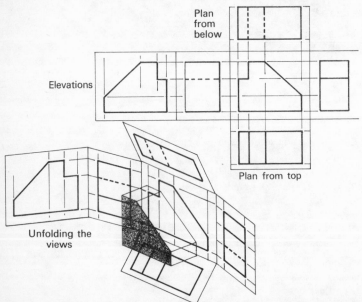

Fig. 6.5. The six views of a First Angle orthographic projection. Notice that all the elevations are in line.

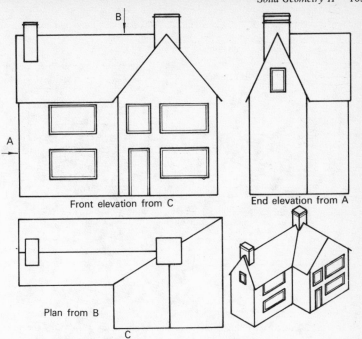

Fig. 6.6. A First Angle projection of a house

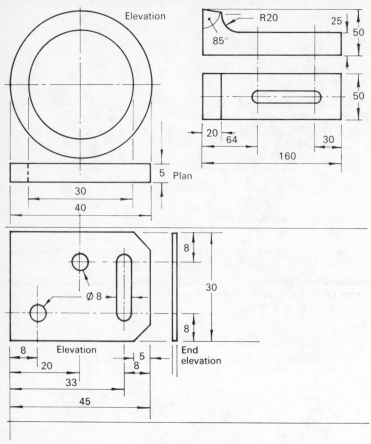

Fig. 6.7.   *Fig. 6.8.* Grinding machine work rest: First Angle projection

Fig. 6.9. Expansion and contraction plate: First Angle projection

The Rabatment or Turning of Solids

It is often necessary to see objects from an angle. To achieve this, we can either move the object, or move ourselves to a new viewpoint. Moving the object is called **Rabatment**. If we move ourselves, we obtain an **Auxiliary View** (see page 110).

Note: Auxiliary means 'one that is helpful'. An extra view.

In Fig. 6.10 a square based pyramid stands on its base ABCD. The examination papers often say, 'lies with its base in the horizontal plane'. The apex lies over the centre of the base. The pyramid is turned round one edge AB so that one face ABO lies in the horizontal plane. Line AB is seen as point A′ in the elevation. We use this point as a hinge pin.

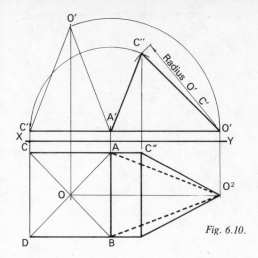

Fig. 6.10.

Method Draw radius A′C′ from centre A′. Draw an arc radius O′C′ from centre O′ to cut the first arc at C″. Complete the elevation and project down to the plan. Remember that C moves parallel to the XY line to C‴.

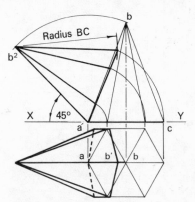

Fig. 6.11.

In Fig. 6.11 a hexagonal based pyramid has been tilted through 45°. The method is similar to Fig. 6.10. Line ab is hinged about a′.

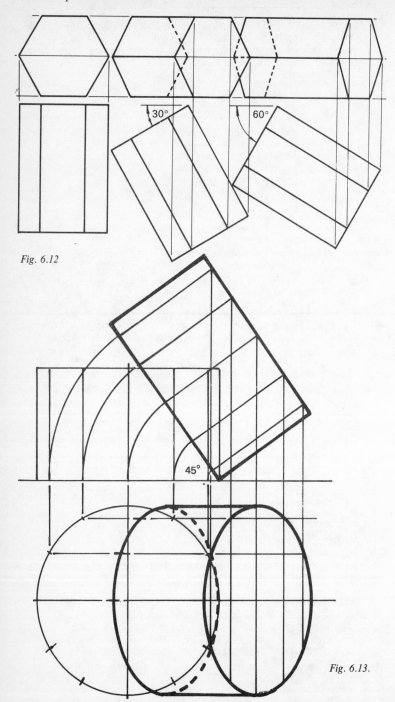

Fig. 6.12

30°   60°

45°

Fig. 6.13.

In Fig. 6.12 a hexagonal prism lies with one face in the horizontal plane. It is then turned **in plan** and the elevations change.

In Fig. 6.13 a cylinder rests with one end in the HP. Tilt it so that the base is inclined at 45° to the HP.

Method Divide the circle into twelve equal parts. Swing the front elevation up as a rectangle. Project each point of the circle and join up.

Fig. 6.12 has corners which can be projected, so there is no reason for dividing the sides. A cylinder has no corners so we must choose points on the circumference. The circle could be divided anywhere but later development work (the folding out of surfaces flat) is much easier when the parts are equal. Dividing circles into twelve equal parts is easy with a 30°/60° set square; is quick and a good habit but it is not compulsory.

In Fig. 6.14 a cone is tilted with its base at 30° to the HP. Turn the front elevation as if it was a simple triangle. Then project the plan and end elevation.

Questions on Rabatment

1 Draw a cube of 50 mm side with one face in the horizontal plane and another

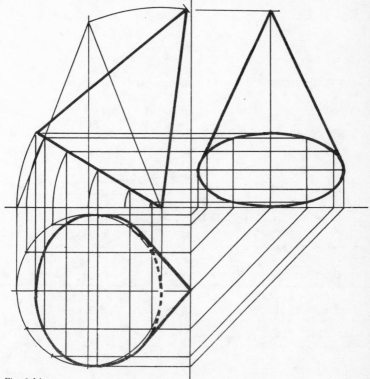

Fig. 6.14.

parallel to the vertical plane. Turn it so that base is inclined at 30° to the HP. (Notice that 'Horizontal Plane' and HP are the same.) Draw the plan.

2 A square based pyramid (of 50 mm side of base and 70 mm vertical height) has one edge parallel to the VP (Vertical Plane). Tilt it so that one face lies in the HP. Draw the plan and the elevation (see Fig. 6.10).

3 Draw Fig. 6.11 with 30 mm side of base and 70 mm vertical height.

4 Draw Fig. 6.12 with a hexagon of 30 mm side and 70 mm long.

5 Draw Fig. 6.13 with a cylinder 30 mm high and 60 mm in diameter.

6 Draw a cylinder 60 mm long and 50 mm in diameter, standing on one end and resting in the HP. Tilt the cylinder so that the base is at 30° to the HP. Draw the plan and two elevations.

7 Draw a cone diameter 50 mm and vertical height 70 mm, with its base tilted at 45° to the HP. Draw three views.

Auxiliary Views

Instead of turning an object, we can move to a new viewpoint and project an auxiliary view. An auxiliary view must *always* be drawn on a plane at right angles to the line of view. If not, you get the same effect as looking at a film from the corner seat in the front row of a cinema. The screen is tilted, so you get a distorted picture which is tall and thin.

A view projected from a plan gives an **elevation.** A view projected from an elevation is a **plan.**

Auxiliary Elevations

In Fig. 6.15 a matchbox is shown in a corner and a vertical plane has been

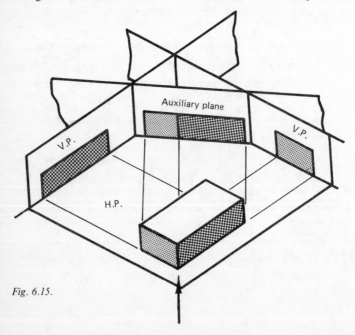

Fig. 6.15.

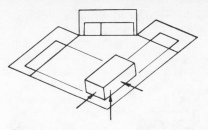

Fig. 6.16.

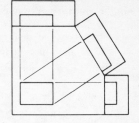

Fig. 6.17.

placed behind it, at right angles to the arrow. The planes have been turned flat in Fig. 6.16 and in Fig. 6.17 the planes have been turned to the normal positions.

In Fig. 6.18 a matchbox is shown in plan and elevation. The elevation is viewed from A. A new elevation is projected from B on a new XY line called X′Y′. The heights remain the same.

A circular disk is shown similarly in Fig. 6.19. Draw the front ellipse abcd by measuring the distances 1, 2 and 3 above the XY lines. Then draw the rear ellipse as follows: draw a series of lines f, g, h, etc. parallel to the X′Y′ line. Project e to e′ on line j. Mark the length e′j′ along each **horizontal** line and join up. This is quicker than drawing another ellipse by projection.

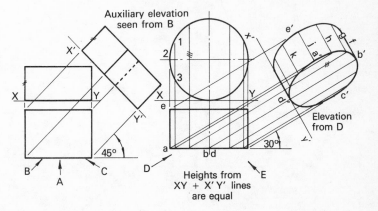

Fig. 6.18. Auxiliary elevation

Fig. 6.19. Auxiliary elevation

We are used to walking round churches and other tall buildings and seeing the roofs 'dance'. As we walk, the roofs appear to move round each other. A model of a church is an excellent visual aid and can be used in many ways. This one consists of a rectangular prism, a square prism, a triangular prism and a square based pyramid. Vertical planes can be moved around behind to show the projection of First Angle auxiliary elevations (see Fig. 6.20).

Fig. 6.21 looks complicated but it is not. Draw the plan and elevation from A to a small size in the middle of the paper. Then change your viewpoint and project as many new elevations as you can get on the paper.

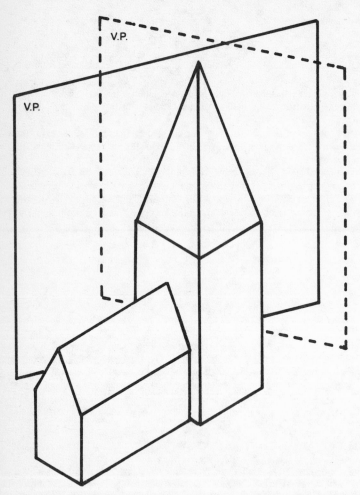

Fig. 6.20. A model of a church with vertical planes behind: Visual aid

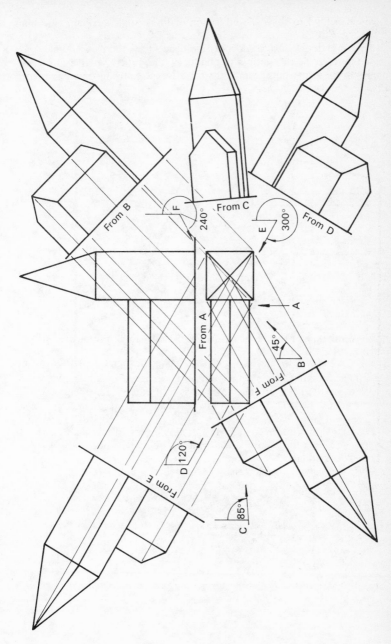

Fig. 6.21.

Auxiliary Plans

Auxiliary plans are projected from **elevations.**

In Fig. 6.22 the elevation is viewed from A and the X′Y′ line is at **right angles** to the projection lines. It is easy to get this wrong. Note that plan thicknesses stay the same.

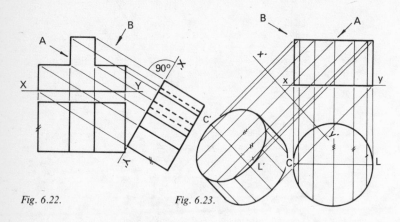

Fig. 6.22.   *Fig. 6.23.*

The centre line C^1L^1 in Fig. 6.23 can be placed at any convenient position. Then take widths from the centre line.

Auxiliary plans are sometimes more difficult to imagine than auxiliary elevations. You have to imagine yourself in an aeroplane or as a worm in a churchyard. The board shown in Fig. 6.24 can be very helpful when people are puzzled. The plan unit pivots round on the top screw and then the plan can be folded flat.

Fig 6.25 shows an imaginary view of the visual aid laid flat. The plan distances (A) are the distances which the church is in front of the vertical plane. The foundations have been tinted.

See how many auxiliary plans you can get on a page (see Fig. 6.26).

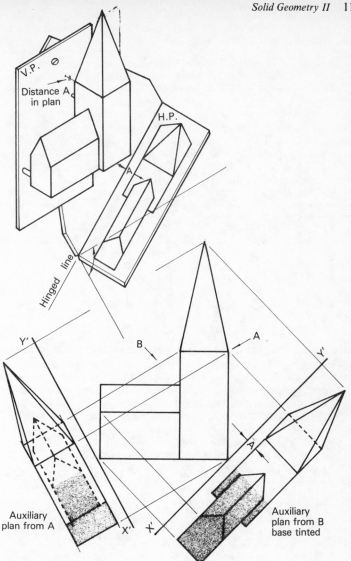

Fig. 6.25.

Fig. 6.24. Auxiliary plans: Visual aid

A solid model of a church fixed away from the vertical plane. A second board with a hinged plan (HP) pivots on the VP.

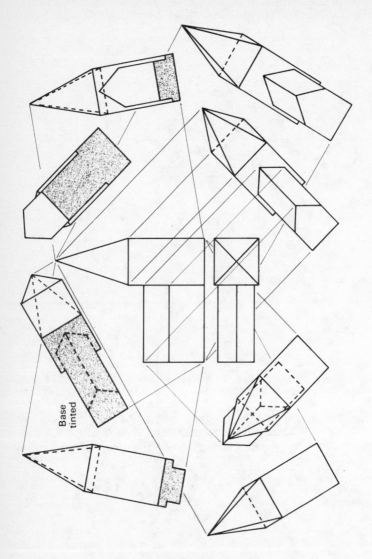

Base
tinted

Fig. 6.26. Auxiliary plans projected from the elevation

Questions on Auxiliary Views

1 Draw Fig. 6.18 double its size. Project an auxiliary elevation from B at 45° and another from C at 30°.
2 Draw Fig. 6.19 double its size. Project auxiliary elevations from D at 30° and E at 45°.
3 Draw Fig. 6.20 to a convenient size. See how many views you can get on the page.
4 Draw Fig. 6.22 and project auxiliary plans from A at 45° and from B at 45°.

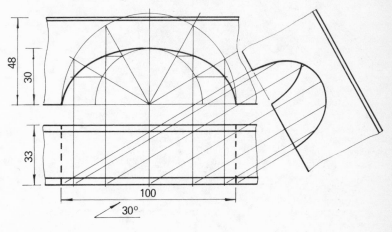

Fig. 6.27.

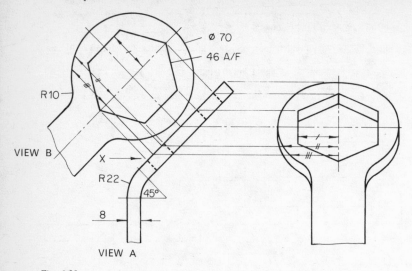

Fig. 6.28.

An auxiliary view is necessary before the ellipse can be drawn

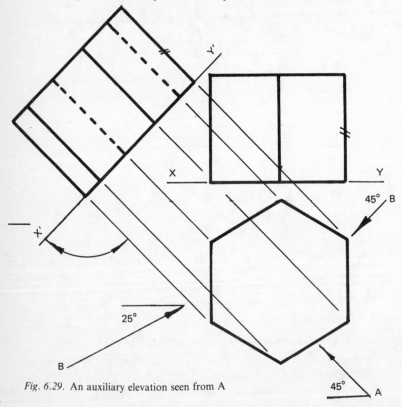

Fig. 6.29. An auxiliary elevation seen from A

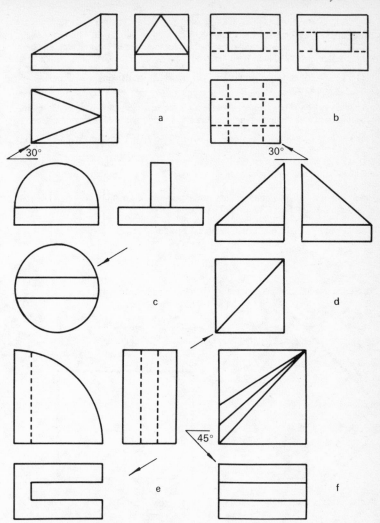

Fig. 6.30. Blocks for auxiliary projection

5 Draw Fig. 6.23 and add auxiliary plans from A at 45° and B at 30°.
6 Draw Fig. 6.26. Place the church in the centre of an A3 sheet of paper with the steeple 85 mm high. Project as many auxiliary plans as will go on the sheet.
7 Fig 6.27 and 6.28 are typical examination type questions. The examiner might not give you any drawing at all but only the following description. Work it out carefully. This is a description of a stone road bridge over a river.

An elliptical arch with major axis 10 metres and half minor axis of 3 metres, is 3.3 metres in total width. The height of the parapet is 4.8 metres. Construct the half ellipse. Draw the front elevation and plan and an auxiliary elevation seen at 30° to the parapet. Add a second auxiliary elevation at 60° from the right. ('Construct' here means that you must use some standard construction **and leave the construction in.**) Scale 1 : 100 (See Fig. 6.27.)

8 Fig. 6.28. The examiner might give you views A and B with measurements and ask the following:

   Two views of a ring spanner are shown. Construct these views showing all points of tangency and the hexagon construction. Then project a side elevation from X.

9 Draw Fig. 6.29 with a hexagon of 40 mm side and 25 mm high. Draw the auxiliary view from A and add another from B.

10 Draw a small object, in front elevation and plan and add an auxiliary view which helps our understanding of the object. This is the sole purpose of an auxiliary view.

11 Some examiners give small articles to be drawn in various projections. Use some of the blocks in Fig. 6.30 for this exercise.

Solid geometry continues in Chapter 8 (page 141).

7
Pictorial Drawing II

Continued from Chapter 4 (page 76)

Isometric Drawing

The Isometric Circle

Fig. 7.1 shows a circle set out inside a square with verticals. The sides of a cube have been set out in Fig. 7.2. with the same verticals. Measure up and down on each vertical. Because the quarter circles are all the same shape the measurement on each vertical is the same top and bottom.

An isometric dice is shown in Fig. 7.3. Draw the three ellipses and then draw the circle as a test of accuracy.

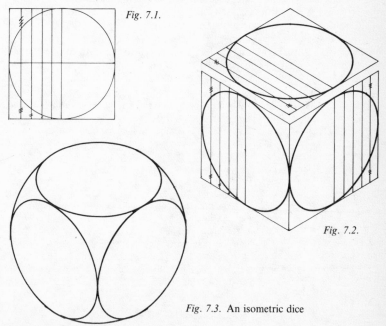

Fig. 7.1.

Fig. 7.2.

Fig. 7.3. An isometric dice

Fig. 7.4. The orthographic projection of a ring with a square section

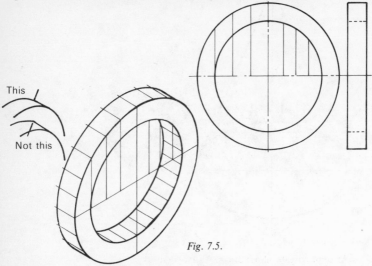

This

Not this

Fig. 7.5.

Fig. 7.4 shows the orthographic projection (First or Third Angle) of a ring with square section. One quarter has been divided for drawing the large circle, and one for drawing the small. Draw centre lines and construct one end face (see Fig. 7.5). Draw lines from the ellipses for marking out the thickness. Mark the thickness along all these lines with one setting of the compass. Take as many points as you need. Join up. Notice that there is no point on the back curve: it must smooth in.

Fig. 7.6 shows a very simple orthographic projection of a right cone. Instead of a full plan, a half plan has been added to the bottom line of the elevation. This method is often used for speed.

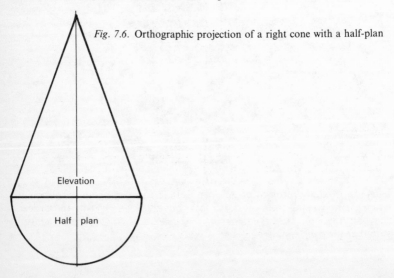

Fig. 7.6. Orthographic projection of a right cone with a half-plan

Elevation

Half | plan

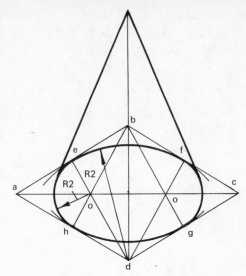

Fig. 7.7 An approximate ellipse drawn with a compass

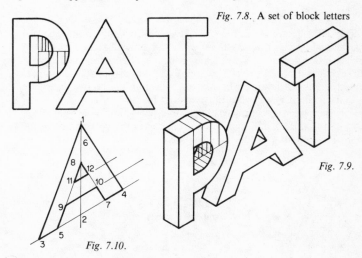

Fig. 7.8. A set of block letters

Fig. 7.9.

Fig. 7.10.

Other Isometric Drawing

In Fig. 7.7 an isometric square abcd has been drawn and an ellipse constructed inside it with compasses. (Construction shown in Fig. 5.29.)

Fig 7.8 shows a set of block letters. Draw the front elevation of the letters in isometric projection (see Fig. 7.9). 'A' is a particularly difficult letter as the sloping lines must be drawn parallel. Draw it in the order shown in Fig. 7.10. Add the thickness to the **nearest letter** (P) first to make sure that the overlapping of P and A is correct.

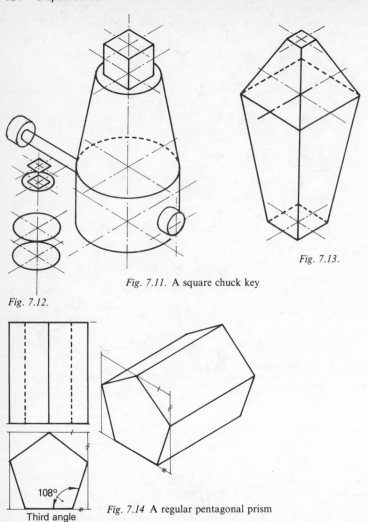

Fig. 7.11. A square chuck key

Fig. 7.12.

Fig. 7.13.

108°

Third angle

Fig. 7.14 A regular pentagonal prism

A Square Lathe Chuck Key

Method (See Fig. 7.11.) Draw a vertical centre line. Draw a series of pairs of centre lines at the correct heights and on them construct squares and circles as if they were plates, one above the other, as in Fig. 7.12.

Construct Fig. 7.13 similarly.

Fig. 7.14 shows a **regular pentagonal prism** (Third Angle) constructed using a protractor. Notice that the angle size has been added. A regular pentagon can be drawn quickly this way but you must print in the angle so that the examiner realises how you have done it. Draw a rectangle round the pentagon. Draw the pentagon inside an isometric rectangle. Add lengths as usual.

A Stepped Block in First Angle Projection

(See Fig. 7.15.) Draw the stepped end as seen from A. We must imagine ourselves in position A and decide if the steps will be on our left or right. Then draw the upright side first and the surfaces of the steps last. Views from A and C are shown. Draw these and the view from B.

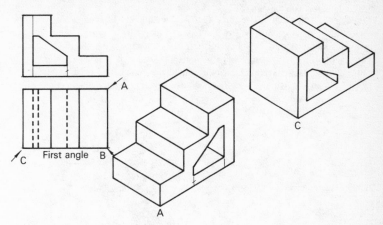

Fig. 7.15. A stepped block in First Angle projection

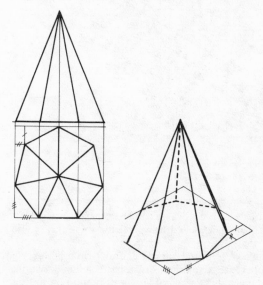

Fig. 7.16. A heptagonal (seven sided) based right (upright) pyramid in First Angle projection

A Heptagonal (Seven Sided) Based Right (Upright) Pyramid Drawn in First Angle Projection

The heptagon in Fig. 7.16 must be drawn with compasses as the corner angle is $128°571^1$, which is impossible using a protractor. Construct a rectangle round the heptagon and draw the pyramid as usual.

Isometric Sketching on Isometric Grid Paper

Isometric grid paper, usually printed in green or brown, is very easy to use. **There is only one rule. Place the paper so that one set of the lines is vertical.** In isometric drawing, uprights lines stay upright. If your paper has no lines standing upright, you cannot draw isometrics on it. In every examination, someone makes this mistake. Be sure it is not you.

Fig. 7.17 shows a stepped block sketched on isometric grid paper.

Figs. 7.18 and 7.19 show cylindrical objects sketched round centre lines.

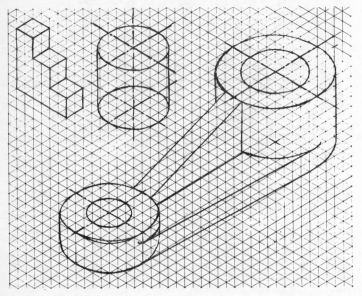

Fig. 7.17. A stepped block drawn on an isometric grid
Fig. 7.18. A cylinder sketched round centre lines

Leave centre lines in
Fig. 7.19. A casting built up of cylinders

Fig 7.20 shows a cast iron angle plate drawn freehand on isometric grid paper.

Fig 7.21 shows the same plate to the same size, drawn on a **hypometric grid.** This grid is based on 3:4:5 triangles so with a basic unit of 3 mm, the vertical lines are 12 mm apart, the horizontal lines are 9 mm apart and the distance between the diagonal intersections is 15 mm.

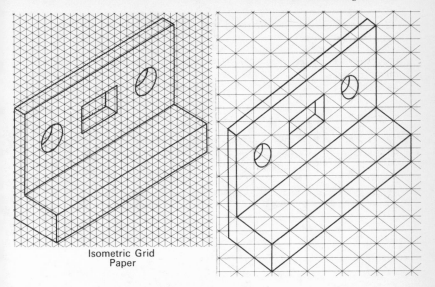

Fig. 7.20. An angle plate drawn on an **isometric grid**

Fig. 7.21. An angle plate drawn on a **hypometric grid**

Visual Aids

The edges of an isometric cube are drawn at 120° to each other. If a real cube is tilted at 35°16′ to the ground and viewed from the front, the corners appear at exactly 120°.

Fig. 7.22 shows a cube resting on an inclined block cut to 35°16′. Any object placed on a block or shelf at this angle, and viewed from the front, will be seen as an isometric view. A shelf sloping at 35°16′ can be of help in a drawing office as one can visualise what is being drawn.

Fig. 7.23 shows an 'isometric slope' holding a cube so that the bottom edges ab (and bc which is unseen) are inclined at 45° to the front edge of the block.

Fig. 7.24 shows the development of a cube with three faces marked out with circles.

Fig. 7.25 shows the isometric cube with the circles seen as ellipses.

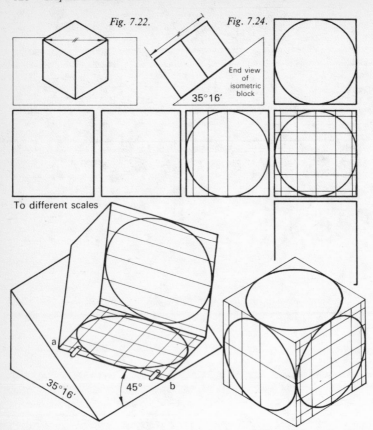

Fig. 7.22.

Fig. 7.24.

35°16′

End view
of
isometric
block

To different scales

a

35°16′

45°   b

Fig. 7.23. The cube on an 'isometric slope'

Fig. 7.25. The cube seen
in isometric

Teaching Aids

Fig. 7.26 shows a solid wooden boat drawn in First Angle orthographic projection. Width and height lines have been scribed at regular intervals along the block. This is a useful teaching aid as it is simple, durable and attractive.

Fig. 7.27 shows the boat on an isometric slope.

Fig. 7.28 is an isometric construction of the boat.

Method Draw the centre line of the base. Draw parallel width lines all along. Measure each width from the centre line and join up to make the base shape. Draw verticals from the ends of all the width lines. Measure the correct height up each vertical and join up the top edges of the boat.

Fig. 7.29 shows an octagonal prism inside a transparent plastic box. This is a teaching aid which demonstrates how many isometric drawings are

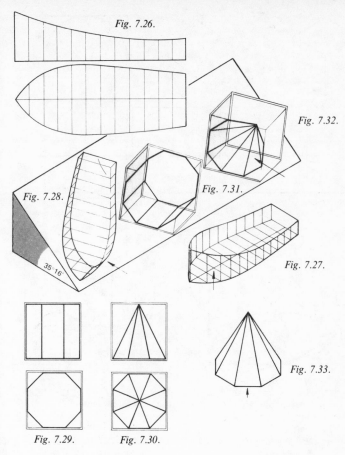

Fig. 7.26.

Fig. 7.32.

Fig. 7.28.

Fig. 7.31.

35° 16'

Fig. 7.27.

Fig. 7.33.

Fig. 7.29.   Fig. 7.30.

Fig. 7.26. A solid wooden boat drawn in First Angle orthographic projection
Fig. 7.27. Teaching aid
Fig. 7.28. An isometric construction of the boat
Fig. 7.29. An octagonal prism inside a transparent plastic box
Fig. 7.30. An octagonal pyramid in a transparent crate
Figs. 7.31 and *7.32.* The crated blocks on an isometric slope
Fig. 7.33. The octagonal pyramid drawn in isometric projection

made by 'crating' an object. The 'crate' which just contains the object, is drawn and then extra pieces are cut away.

It is a lot easier, when making these crated visual aid blocks, to find the plastic boxes first and make the blocks to fit them.

Figs. 7.30–7.33 are self-explanatory.

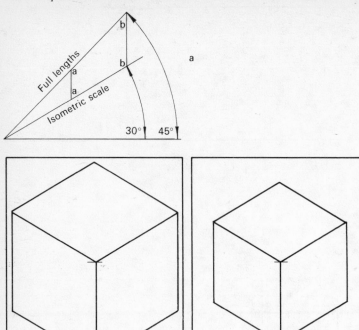

Fig. 7.34. (*a*) An isometric scale
       (*b*) Full size isometric drawing of a cube
       (*c*) The same cube drawn using isometric scale on all three axes

Isometric Scale

Isometric drawings tend to look bigger than they should. The eye expects lines sloping away along the axes, to appear smaller. An isometric scale makes the size appear more realistic. Fig. 7.34(*a*) shows an isometric scale. True lengths are measured along the 45° line and projected down to the 30° one as at aa, bb. The two cubes (see Fig. 7.34(*b*) and (*c*)) show the different results.

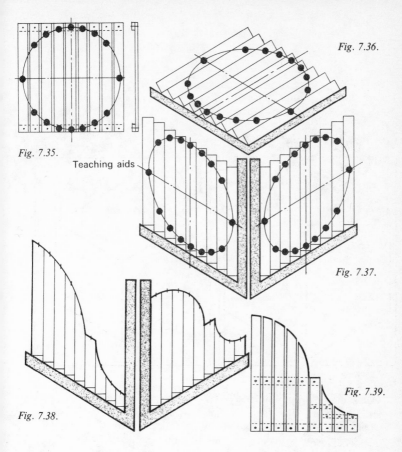

Fig. 7.36.

Fig. 7.35.

Teaching aids

Fig. 7.37.

Fig. 7.38.

Fig. 7.39.

Examination questions are seldom set using isometric scale until 'A' level.

Fig. 7.35 shows a mat made of slats loosely screwed to two back strips. A circle has been scribed on the front.

In Fig. 7.36 there are two wooden ledges at 120° to each other and the mat is resting in the angle. The **dots** marked on the centre lines of the slats have moved into the shape of an ellipse.

Fig. 7.37 shows two similar sets of ledges at 60° to each other, illustrating the construction of vertical isometric ellipses.

Figs. 7.38 and 7.39 show another mat. The quarter ellipses have been drawn through the points as before.

Fig. 7.40. A barrel vault crossing at Athos from *The Art of Roman Building* by A. Choissy

An isometric projection has been redrawn in Fig. 7.40 from *Art de Batir Chez les Romans* (*The Art of Roman Building*) by A. Choissy, depicting a barrel vault crossing at Athos. Modern architectural tints have been used and scraped away to give form. This, and Figs. 7.41 and 7.44, of Brunel's Bridge and a three-jaw chuck, are more advanced work which could be called Technical Illustration. Such drawings take time and care, but are just as easy to understand as the earlier work.

Fig. 7.41 shows Brunel's Clifton Suspension Bridge. The drawings have been shaded with architectural tints. Some of the rocks have double layers and have been scraped away with a knife. The drawing could equally well be shaded in coloured pencil or washed with water colour. One half of the elevation is shown in Fig. 7.42. This was made by careful measurement of a photograph. Then Fig. 7.41 was made by measuring the horizontal gaps between the vertical cables and their lengths. It was much simpler than it looks. Compare with Fig. 4.14.

Fig. 7.43 shows the buttress end which the shareholders made Brunel build. He wanted to start the bridge further inland, thus making the buttress unnecessary, but they said that the suspension bridge would then be too long and therefore dangerous. The buttress cost so much to build that the venture went bankrupt and the bridge was not completed until after Brunel's death, years later.

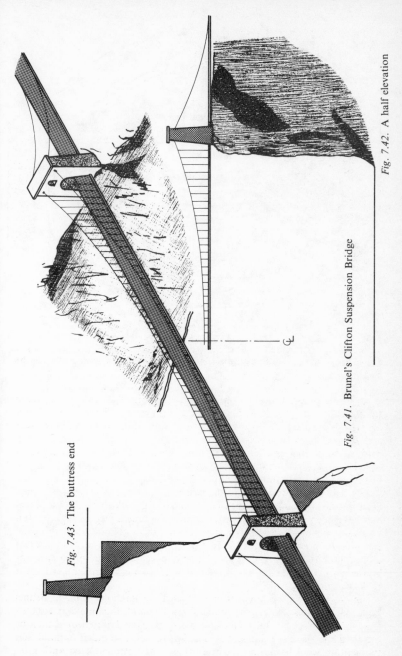

Fig. 7.43. The buttress end

Fig. 7.42. A half elevation

Fig. 7.41. Brunel's Clifton Suspension Bridge

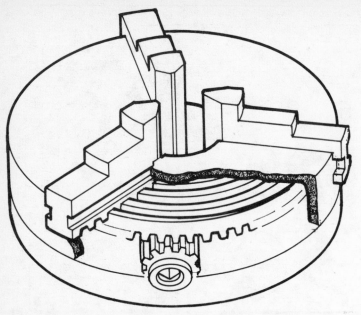

Fig. 7.44. A cut-away isometric projection of a three-jaw chuck

A cut-away drawing of a three-jaw chuck is shown in Fig. 7.44. This has been drawn with elliptical plastic templates. The edges which are bordered by space have been thickened. Ones where two surfaces meet have been left thin. This technique makes a drawing more dramatic. The broken surfaces have been tinted. This is very advanced work but has been included for interest.

Oblique Projection

Oblique projection is probably the most simple form of 'picture drawing'.

Fig. 7.45 is a good example of oblique projection. The front elevation of the letter P has been drawn with instruments, the centre of the semicircle projected back at 45° and the thickness measured back to full size. This is the centre of the new back semicircle.

The rules for oblique projection are:- draw the front surface to the true shape; draw thickness lines back at 45°; and measure all thicknesses to full size.

7.46 shows the front elevation and plan, in First Angle orthographic projection, of a stepped block. The front surface of the front block has been shaded. The front surface of the bottom block has been drawn true shape in Fig. 7.47. Thickness lines have been projected at 45° and measured at full length. Points a and b have been found by measuring and the front surface of the second step drawn to true shape. The process has been repeated.

Fig. 7.45.

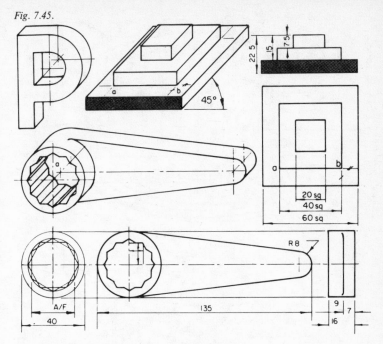

Fig. 7.46. A stepped block in First Angle projection
Fig. 7.47. A stepped block in oblique projection
Fig. 7.48. A box spanner in Third Angle projection
Fig. 7.49. A box spanner in oblique projection

Fig. 7.48 depicts a box spanner in Third Angle projection. The construction of the double hexagon has been drawn by the side to avoid confusing the drawing. The oblique projection (see Fig. 7.49) needs thought.

Method Draw the front surface of the spanner, with its double hexagons, on the centre lines. Project back the centre point. Mark the centre points 9

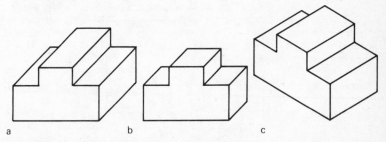

a   b   c

Fig. 7.50. (a) Cavalier projection – oblique projection without reduction
(b) Cabinet projection – oblique projection with sloping lengths reduced by half
(c) Isometric projection – with lengths true size

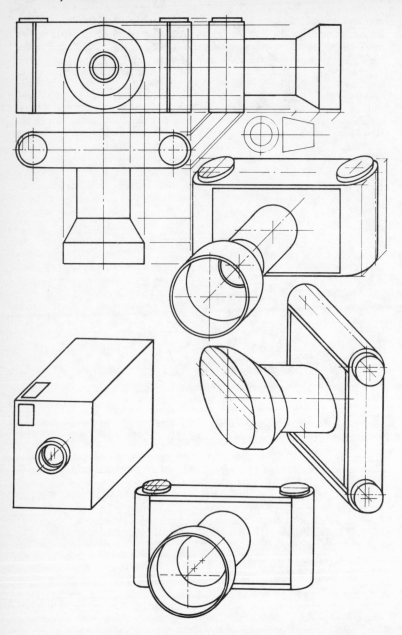

mm and 16 mm back for the other part-circles. From the centre a, draw the centre line for the handle. Draw the inside edges of the spanner by measuring back. Do not draw circles. Draw the sloping sides of the handle with a ruler. There is no need for tangency construction.

The three views of the same block in Fig. 7.50(*a*), (*b*) and (*c*) show the different types of picture drawing. Oblique projection tends to exaggerate the thickness. Cabinet projection tries to improve the appearance by halving the thickness lengths. Isometric projection tilts both width and thickness lines away at the same angle so the problem does not arise.

Exercises

Draw all these examples in oblique projection, cabinet projection and isometric projection. Label **every** projection so that you learn them.

Figs 7.51–7.56 have been drawn to illustrate good and poor uses of oblique projection. Fig. 7.51 works well because the only circles are on the front face and so can be drawn with a compass. We do not know how long the camera box is because we do not known if it has been shortened. Advertisers use this as a trick. They draw bars of chocolate as if they are a mile long.

Fig. 7.52 is a good, straightforward orthographic projection.

Fig 7.53 looks as if the camera has a telephoto lens because the thickness

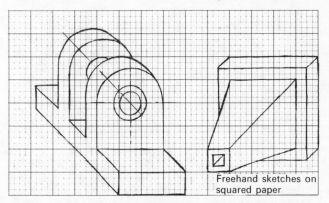

Freehand sketches on squared paper

Fig. 7.56. Freehand sketch on squared paper of an oblique projection
Fig. 7.57. Freehand sketch on squared paper of an oblique projection

Fig. 7.51. A box camera in oblique projection
Note: Try to arrange that the circles are seen as circles. They are badly distorted when seen otherwise.
Fig. 7.52. First Angle orthographic projection
Fig. 7.53. Cavalier projection of a camera
Fig. 7.54. Cabinet projection of a camera
Fig. 7.55. Cavalier projection

has not been reduced. This is a cavalier projection. Fig. 7.54 is the same drawing in cabinet projection and, because thicknesses have been halved, it looks more accurate.

In Fig. 7.55 drawn in cavalier projection again, the camera body looks very wide.

Oblique projections are easy to draw freehand on square grid paper (see Figs 7.56 and 7.57). Always try to draw the circles true shape. If this is not possible, then use another projection. Oblique projection is not suitable.

Problem

Find an object with circles on one face only. It must not be too complicated. Make a freehand sketch of it on squared paper. **From this sketch**, make an oblique projection using instruments. Check with the object that your drawing is correct. If not, trace back through your work and see where you went wrong.

Pictorial drawing continues in Chapter 9 (page 165).

8
Developments – Solid Geometry III

Continued from Chapter 6 (page 122)

Cardboard boxes are cut out flat and folded up. Collect as many small boxes as you can of every shape, and unfold them. This will teach you more about developments than a long lecture. The art of packaging is very sophisticated and practitioners call themselves Cardboard Engineers.

Visual Aid

Make a collection of pairs of boxes, one folded and the other laid out flat. Display them on a board to show the self-locking tabs, the cut-out inserts to hold bottles, etc. safely, and all the variety of shapes that can be formed from a flat sheet. Add a few sketches on the lines of Fig. 8.1.

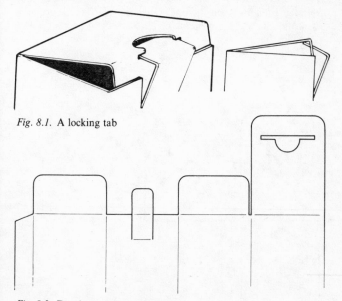

Fig. 8.1. A locking tab

Fig. 8.2. Development of a block

Fig. 8.2 shows a stepped box laid out flat. Each surface folds out **at right angles to the hinge line.**

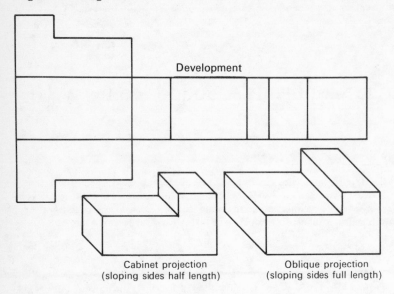

Development

Cabinet projection
(sloping sides half length)

Oblique projection
(sloping sides full length)

Fig. 8.3 shows a hexagonal box with a sloping lid. The true shape of the lid has been found by hinging it on the line. Distances from the hinge are measured as shown by the arrows in the plan. The developed box is shown in Fig. 8.4. The sides have been folded out and heights taken from the front elevation. The base is a regular hexagon and the lid is shape A.

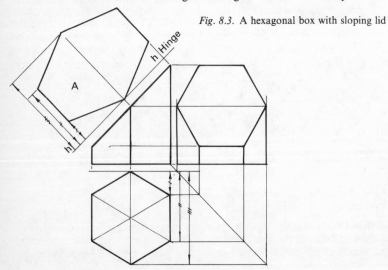

Fig. 8.3. A hexagonal box with sloping lid

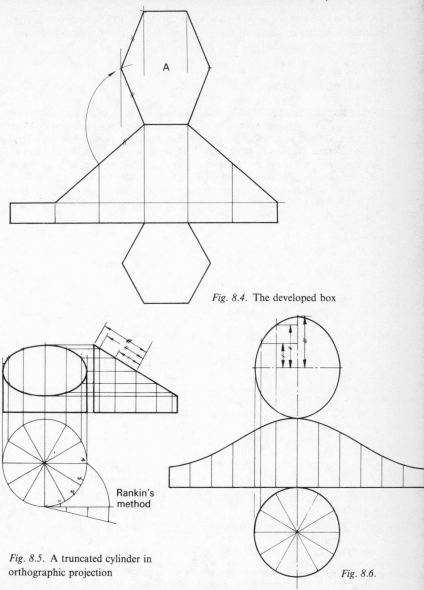

Fig. 8.4. The developed box

Rankin's method

Fig. 8.5. A truncated cylinder in orthographic projection

Fig. 8.6.

Fig. 8.5 shows a cylindrical box with a sloping end Fig. 8.6 shows it developed.

Method Divide the plan into twelve equal parts. Dividing into equal parts saves time because the measurements for each quarter are the same and one setting of the compass can be used four times. Find the circumference by Rankin's method, or by calculation, or by the method shown in Fig. 2.65. Find the true shape of the sloping end as shown by the arrows and by projecting from the plan (see Fig. 8.6).

True Lengths of Lines

It is important to be able to recognise which lines are seen as true lengths and which lines are foreshortened. **Lines are seen as true lengths only when they are parallel to the vertical and/or the horizontal plane or an auxiliary plane.** Before going any further with developments, we must study **true lengths** and realise when we are looking at lengths which have been shortened by being turned away from us.

Fig. 8.7 shows two set squares being turned and viewed in plan and elevation. The true lengths of side are shown in dark lines. All the other lines are shorter than true length.

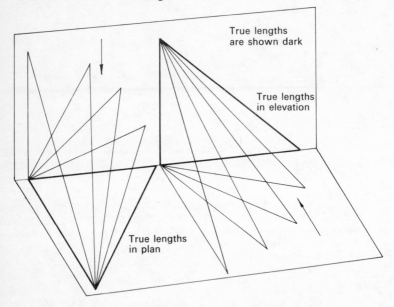

Fig. 8.7. Set squares turned to illustrate true and false lengths of lines

The diagrams in Fig. 8.8(*a*) and (*b*) show lines in the corner where a vertical and a horizontal plane meet. The same lines are drawn in orthographic projection (see Fig. 8.8(*b*)).

Line 1 is parallel to the VP and to the HP. Therefore it is shown as a dark line in both plan and elevation as they are both true lengths.
Line 2 is parallel to the VP so the elevation is a true length.
Line 3 is parallel to the HP so the plan is a true length.

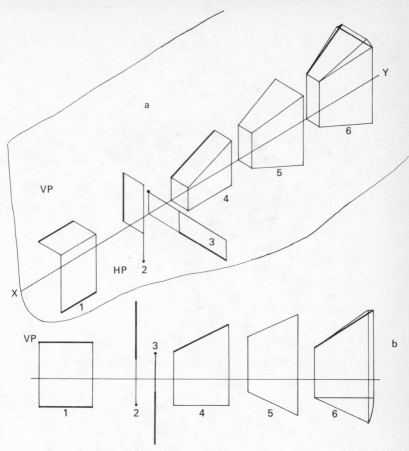

Fig. 8.8. True lengths of lines

    (*a*) Isometric projections of lines in space

    (*b*) Orthographic projections of the same lines

Note: True lengths are shown as dark lines

Line 4 is parallel to the VP so the elevation is a true length.

Line 5 is inclined to the VP and to the HP **so no view is true length.**

Line 6 has been turned in plan so that it is parallel to the VP. **This has made the new elevation of line 6 into a true length.**

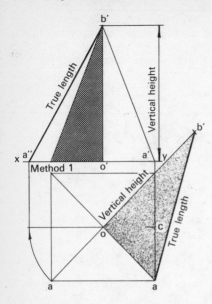

Fig. 8.9. Finding the true length of a sloping corner

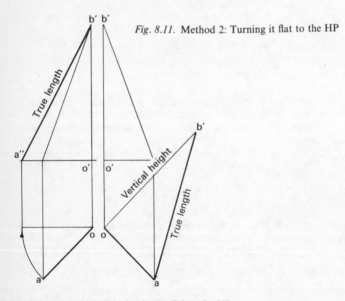

Fig. 8.11. Method 2: Turning it flat to the HP

Fig. 8.10. Method 1: Turning it parallel to the VP

On a square based pyramid (see Fig. 8.9), only the base is seen as a true shape. The sloping corners have to be turned if we are to see them as true lengths. The line a′b′ is the true length of the centre line bc, not of ba. The shaded section in Fig. 8.9 is **an imaginary internal triangle.** It is Oab in plan.

To Find the True Length of Line ab

With centre O and radius Oa, turn line Oa until it is parallel to the xy line. Project to draw a″b′. This is the true length of the line. Fig. 8.10 shows the internal triangle by itself, turned to show the true length.

Second Method of Finding the True Length of a Line ab

In Fig. 8.11 Oab′ (the internal triangle) has been laid flat. It has been hinged along Oa so that angle b′Oa is 90° and b′O is the vertical height of the pyramid.

Visual Aid to Show an Internal Triangle in a Square Based Pyramid

A wooden, square based pyramid has a saw-cut along one corner through to the centre of the base (see Fig. 8.12). A thin piece of cardboard is inserted into the cut and the imaginary internal triangle a′b′o′ is marked. The card can then be removed and revolved to show the movement of an 'internal triangle'. This can be a useful teaching aid as many students find the construction difficult to visualise. The triangle and rod can be made in metal and stored in the slot.

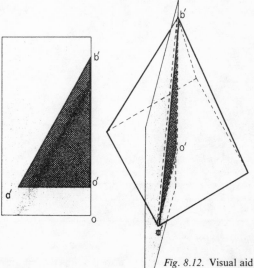

Fig. 8.12. Visual aid to show an internal triangle
A solid pyramid with a saw-cut and paper inserted

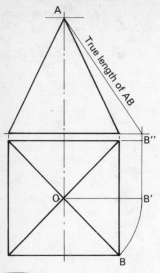

Fig. 8.13. Development of a square based pyramid

Developments

The Development of a Square Based Pyramid

Using the 'internal triangle' construction, we can now develop the square based pyramid (see Fig. 8.13). Find the true length of the sloping edge. Set out an arc of this radius and step off the length of the base four times. Add the square base.

It is difficult to develop a truncated pyramid without completing the pyramid (see Fig. 8.14). Complete the pyramid. Find the true length of the sloping edge and develop as usual.

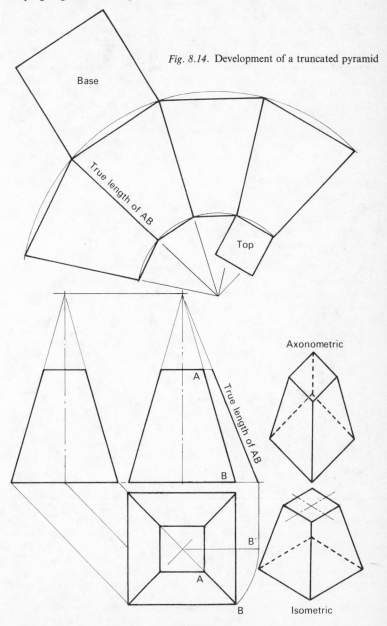

Fig. 8.14. Development of a truncated pyramid

The Development of a Regular Tetrahedron

(See Fig. 8.15.) Find the true length of the sloping edge by turning the internal triangle and develop as before.

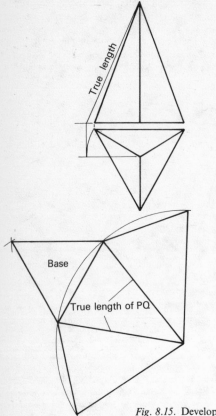

Fig. 8.15. Development of a tetrahedron

A Truncated Cone

The truncated cone must be completed and the development revolved from
the apex (see Fig. 8.16). Step off the chord from the plan as with the
chords in the development. Always start the development slightly away
from the front elevation or the drawing becomes difficult to understand.

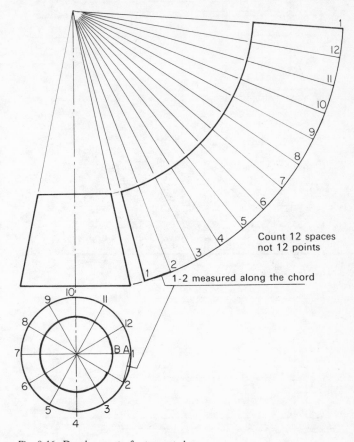

Count 12 spaces
not 12 points

1-2 measured along the chord

Fig. 8.16. Development of a truncated cone

Fig. 8.17. Development of a cut cylinder

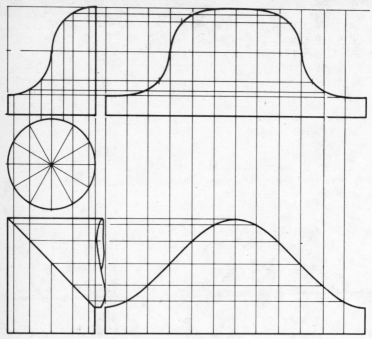

Fig. 8.18. Development of a 90° bend in a pipe

Fig. 8.19. More cylinders

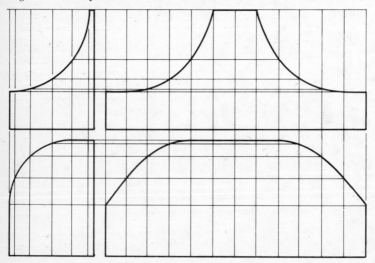

Fig. 8.20. More cylinders

The developments for Figs. 8.17–8.21 are all similar. Here one must draw or calculate the circumference. **Do not step off the chord from the plan as this will be too inaccurate.** The top and bottom curves are plotted point by point and carefully joined. The two sides must be similar shapes. The drawing of these curves needs a lot of practice.

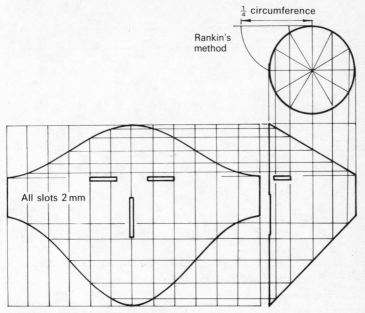

Fig. 8.21. A cylindrical paper mask

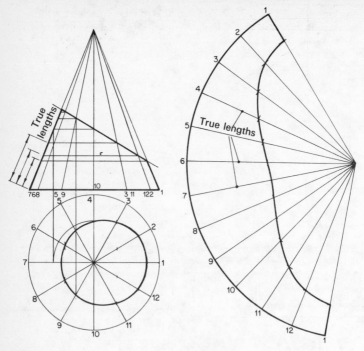

Fig. 8.22.

The development shown in Fig. 8.22 is the examiner's favourite. Make sure you get it right. The trouble is that *all* the sloping lines are false lengths except the two outside ones.

Theory If the cone was cut off level, parallel to the base, all the sloping sides would really be the same length. We would *see* only the two outside ones as true lengths. All the rest would be sloping away from us and we would see them shorter than true length. Try this by holding up a pencil and moving it about like the lines on the cone. We use this to solve the question. **Project all the sloping lines to the side and measure the true lengths there.**

Remember that the examiner marks with the answer drawn on a piece of tracing paper. He lays his correct solution on top of your answer so that any false lengths show up.

A Hexagonal Tray

Method Look along the hinge line and fold the sloping side down (see Fig. 8.23).

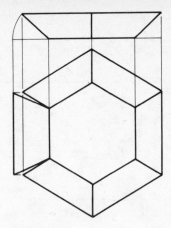

Fig. 8.23. A hexagonal tray

A Hopper

This is slightly more complicated as the sides are of different widths (see Fig. 8.24). We can draw the ends as before, but the front and back developments present problems. Line ab is the true length of a sloping corner. Swing it round on to the line pp to give the width.

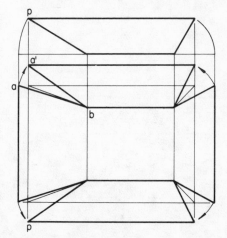

Fig. 8.24. A rectangular hopper

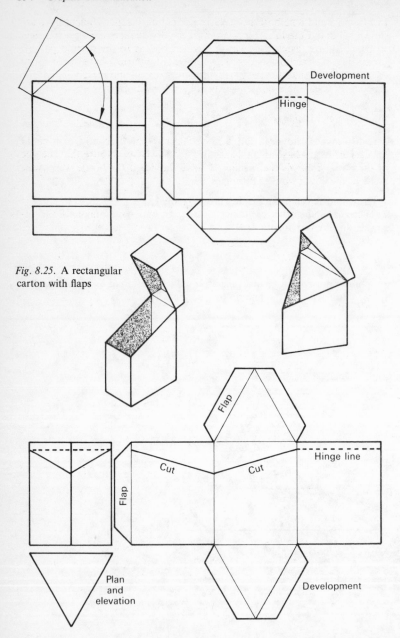

Fig. 8.25. A rectangular carton with flaps

Fig. 8.26. A triangular prism made as a box

In Figs 8.25 and 8.26 both boxes have lids cut at a slope. The cut lines are shown solid and the hinge lines are broken. Box makers have special steel strips which are fastened down on wooden boards to the desired pattern for cutting out the box shapes. The cutting rules (strips) have razor sharp edges, while the rules for the hinge lines are rounded and merely indent the cardboard. Examine a box and you will see the difference.

'Point-of-Sale' Displays

The display box shown in Fig. 8.27 could be in coloured card or it could have a cardboard base and a transparent plastic cover. Notice that the size of the bottle controls the turning of the pyramid. If the bottle were twice as high, the cover would not hinge.

1 Draw and develop the display box as shown.
2 Then find some other container – or design one – and make a Point-of-Sale display box for it.

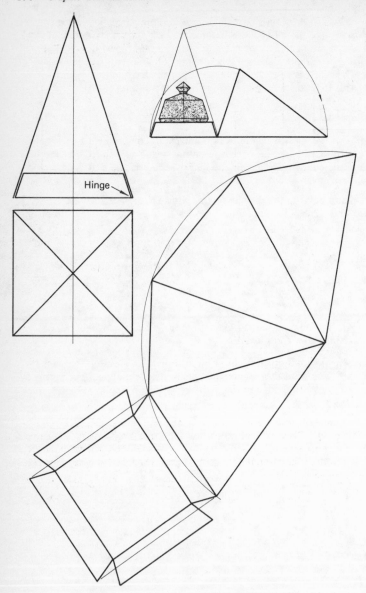

Hinge

Fig. 8.27. A bottle of perfume in a pyramid shaped display box

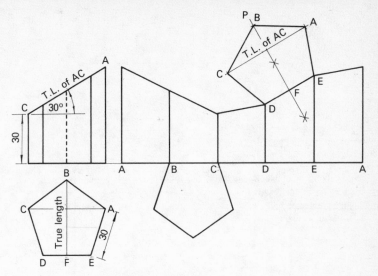

Fig. 8.28. A pentagonal prism cut obliquely

A Pentagonal Prism Cut Obliquely

A regular pentagonal prism is cut at 30° in Fig. 8.28. Develop the sides by folding out. The base is a regular pentagon. The sloping surface is more difficult to draw as no edge in the plan shows a true length.

Method In the plan bisect DE at F. Then FB is a true length, so we can pivot the pentagon round FB. In the development, bisect DE at F. Draw FB at 90° to DE. The end elevation shows the true length of AC. Set out half AC on either side of FB. Join up. We can often find a true shape by pivoting it about a horizontal or a vertical line. **This line (as here) does not have to be an edge.**

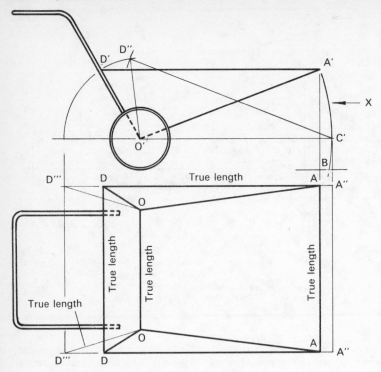

Fig. 8.29. Plan and elevation of a garden barrow

A Garden Barrow

The plan and elevation of a special garden barrow designed for the easy tipping of leaves, etc. is shown in Fig. 8.29. The barrow is shown in isometric in Fig. 8.31.

Question 1

1 Draw the given plan and elevation. Scale off from the drawing to double size.
2 Draw the end elevation from X.
3 Tip the barrow so that the front edge touches the ground at B. Redraw the end elevation and plan with the barrow in this position. Draw the handle curves freehand.

Question 2

1 Redraw the given plan and elevation as in Question 1.1 (see above).
2 Develop the barrow.

This development is difficult because **not one surface is seen as a true shape.** However, lines AD, AA, DD and OO are all true lengths.

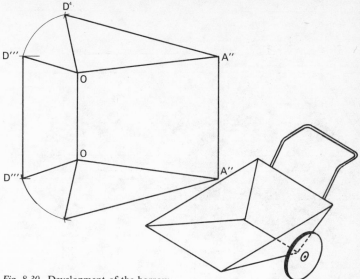

Fig. 8.30. Development of the barrow
Fig. 8.31. Isometric drawing of the barrow

Method (Shown by fine lines in Fig. 8.29). Tilt the barrow until A′ is level with the axle at C′. In plan, A moves to A″ and OOA″A″ is a true shape. Set out this true shape away from the drawing in Fig. 8.30. In the elevation of Fig. 8.29, turn D′O′ round O′ until it becomes horizontal at D″. In the plan, D moves out at right angles to the hinge line OO. Project down from D‴ to D″. Then the line OD‴ is a true length. Add the shape OOD″D‴ to Fig. 8.30.

We now have all the true lengths of the edges of the barrow and can turn side ODA flat. With centre A and radius equal to true length AD, draw an arc. With centre O and radius OD″ cut the previous arc at D⁴. Complete the development.

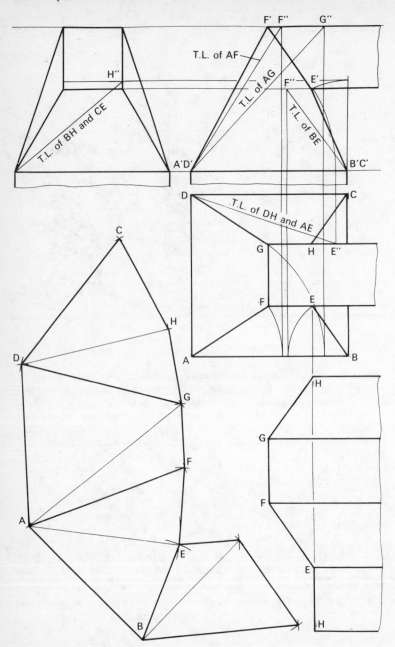

Fig. 8.32. The development of a forge hood

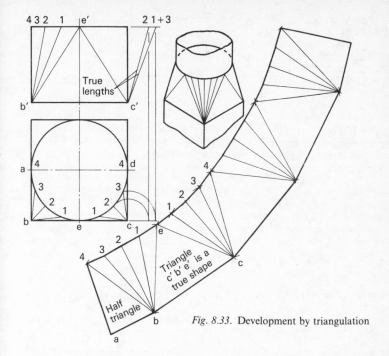

Fig. 8.33. Development by triangulation

The Development of a Forge Hood

The forge hood is made of two sheets of mild steel. Develop the horizontal section by folding out. This is simple (see Fig. 8.32). The pyramid section is developed as a series of triangles. Start the development by drawing AB which is a true length. Find the true length of BE by swinging round in plan and projecting up to E″. Find the true length of AF by swinging the plan down and projecting to F″. Use these true lengths to draw triangle ABE in the development. Continue to find true lengths and add triangles to the development as you go.

Development by Triangulation

Curved surfaces like cylinders, which are folded in one direction, can be unrolled flat, as a series of rectangles. Cones are unrolled as sets of triangles. More irregular surfaces are unrolled as networks of linked triangles, some one way up and some the other.

Fig. 8.33 shows a **transition piece**, a sheet of metal which joins a square tube and a cylinder. It has been divided into a series of triangles. Triangle b′c′e′ is a true shape. The next four triangles, ce1, c12, etc. are not true shapes. The true lengths of c1, c2 and c3 have been drawn by rabatment, next to the elevation. The development has been drawn as a series of **triangles with true length sides**, which wander across the page as they may.

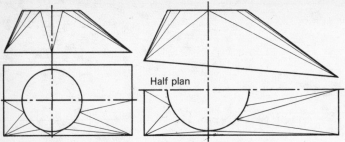

Half plan

Fig. 8.34. Transition pieces from rectangles to circles

Fig. 8.35. Transition pieces from rectangles to circles

Questions

Draw orthographic projections of the following items to double size or larger, unless otherwise stated, and develop them.

1 An L shaped block like Fig. 8.1.
2 A truncated hexagonal prism like Fig. 8.3 with side of base 25 mm. Trace on to thin card, cut out and fold up. How accurately does the lid fit?
3 A truncated cylinder of 50 mm diameter, maximum height 75 mm, cut at 30°. Develop and fold up.
4 A square based pyramid of 50 mm side of base and 80 mm sloping corner. (**Take care** – this is a typical examiner's trick. It is not exactly the same as Fig. 8.13 in construction.)
5 A truncated square based pyramid of 50 mm side of base and 80 mm vertical height. Develop, fold up and draw freehand on an isometric grid. Draw also in axonometric projection on a piece of square grid paper.
6 A truncated cone 50 mm diameter base, 70 mm vertical height, is truncated horizontally to 50 mm vertical height.
7 A cylinder 50 mm dia., 100 mm high which is cut off at 30° at the top and 45° at the bottom.
8 A cone 60 mm diameter base, 75 mm vertical height, which is truncated at 30°. Maximum height 35 mm.
9 Design a hexagonal cardboard box to take a round bottle of ink 50 mm diameter and 80 mm high, comfortably. Show how it will open and develop it.
10 Draw a diagram for advertising the box of ink in any pictorial projection and letter it attractively.
11 Design and make a toy theatre (maximum size 150 mm long). It must be made of thin card, lock in an open position and fold flat for easy storage. It does not have to be a box. It might open like a book or a fan.
12 Draw the plan and elevation of Fig. 8.33 by scaling to double size from the book. Develop the transition piece by triangulation.
13 Repeat the operation for Fig. 8.34, constructing all the true lengths carefully. All construction workings must be shown.
14 Repeat again for Fig. 8.35.
15 Prick through the solutions to questions (Figs. 8.33, 8.34 and 8.35) with a compass point on to a piece of thin card, fold up and make sketches from different directions.

Solid geometry continues in Chapter 10 (page 172).

9
Pictorial Drawing III – Perspective

Continued from Chapter 7 (page 140)

Perspective is a system of drawing used to make things appear natural – to appear as we see them. Other projections are useful but they never look quite right. The rules of perspective can help us to make things look solid and real.

The Greeks experimented with perspective; the Romans wrote on it; but it was not until the Italian Renaissance, in the fourteenth and fifteenth centuries, that the rules of perspective were worked out. At this time perspective drawing exploded. This is one of the reasons why paintings of this period are so exciting. There was a ferment of ideas, with each artist trying to outdo his fellows in getting the perspective right. They particularly relished drawing crowds of people ascending to heaven, or being cast down into hell, because these paintings gave great opportunities for experimenting with perspective: making the crowds gasp.

Pictures became dramatic and even violent, in a way which had never been possible before. At the same time stage scenery became elaborate. Theatres were built long and narrow instead of round, so that the audience could see the elaborate perspectives. Scenery designers developed long vistas. Painters varied their viewpoints, gazed up or down, and explored all the novelties opened up by the new rules of perspective.

Artists added to the sense of reality in their pictures in a second way. They painted the distance in a series of lighter and lighter tones. The nearest mountain was a dark blue, the next range a lighter blue and the furthest, lighter still. The logo of Ultimate, shown as Fig. 18.9, is shaded in a similar way but different tones of architectural tints have had to be substituted for the firm's tones of blue.

This chapter outlines the basic rules of drawing in perspective. Think of the subject as the Italian artists thought of it, as an exciting puzzle to be solved.

One Point Perspective

The engraving in Fig. 9.1 shows the basic rules of **one point perspective**. The man with his back to us is surveying the scene. The horizon is at his eye-level. He is looking straight forward and all backwards and forwards

Fig. 9.1. The principle of one point perspective

lines converge (meet at) the centre of his vision. The sunken courtyard is divided into perspective 'squares'. This was a favourite method used by artists. They drew their floors, often in black and white square tiles, and then knew where to put their table legs and the feet of near or distant people. In the engraving, the corners of the square stones radiate to two vanishing points on the horizon, just outside the picture. Only a few radiating lines are shown but all the corners obey the rule.

Method of Setting Out a Floor

Following Fig. 9.2, divide the nearest line aa into seven equal parts. Radiate to the centre vanishing point. Decide the end of the squared floor bb'. Draw from a through b' to the left vanishing point (LVP) on the horizon. Radiate back from the LVP to all the points on the front edge aa. Draw the squares from 1, 2, 3, etc. parallel to the front edge.

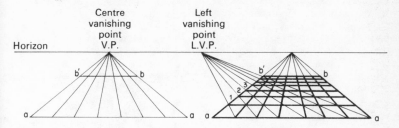

Fig. 9.2. Method of setting out a floor

Single Point Perspective of a Room

At the top of Fig. 9.3 is the plan of a room with a **picture plane** (a sheet of glass held vertically) in front. The perspective view is thought of as being drawn on the picture plane. Take a **station point** (SP) about $2\frac{1}{2}$ times the depth of the room in front of the picture plane. This is where the viewer is stationed to look at the room through the picture plane. Radiate from the station point to the room. Project down from where the radiating lines cut the picture plane. This gives the widths of the picture.

Draw the outside frame of the picture to the correct height abcd. Draw the eye-line. It is interesting to draw the same room from different people's eye levels. A child would never see the counter tops while a giant would see almost a plan. Project down from the SP to the eye-line. This is the vanishing point where all the receding lines meet. Here, all the lines meet in the garden, at a point seen through the window.

In Fig. 9.4 the picture plane has been moved into the room – we see less of the room and more of the counter. The window is larger because it is nearer.

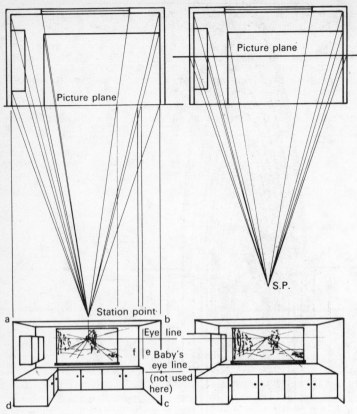

Fig. 9.3. Single point perspective of a room   *Fig. 9.4.*

A one point perspective, in Fig. 9.5, is redrawn from a photograph of the parabolic arches from Casa Battló, built by Gaudi. Gaudi, the Spanish architect, used parabolas a great deal, believing them to be the strongest and most economical of all building shapes. Here, they make a very dramatic picture.

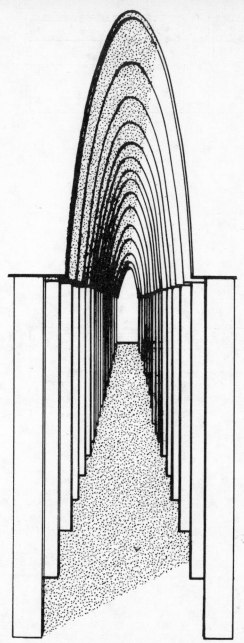

Fig. 9.5. One point perspective from Gaudi's Casa Battló

168   *Graphic Communication*

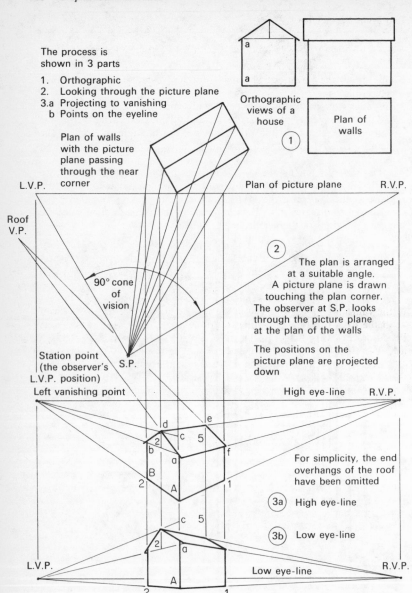

The process is
shown in 3 parts

1. Orthographic
2. Looking through the picture plane
3.a Projecting to vanishing
   b Points on the eyeline

Plan of walls
with the picture
plane passing
through the near
corner

L.V.P.

Roof
V.P.

90° cone
of
vision

Station point
(the observer's
L.V.P. position)
Left vanishing point

Orthographic
views of a
house

Plan of
walls

Plan of picture plane   R.V.P.

② The plan is arranged
at a suitable angle.
A picture plane is drawn
touching the plan corner.
The observer at S.P. looks
through the picture plane
at the plan of the walls

The positions on the
picture plane are projected
down

S.P.

High eye-line   R.V.P.

For simplicity, the end
overhangs of the roof
have been omitted

(3a) High eye-line

(3b) Low eye-line

L.V.P.   R.V.P.
Low eye-line

Fig. 9.6. Two point perspective

Two Point Perspective

Fig. 9.6 describes the basic principles of two point perspective in three parts.

1 This shows the elevations and plan of a simple house in orthographic projection.
2 **This shows the method of drawing.** Arrange the plan at a suitable angle and draw the picture plane to pass through the nearest corner. Choose a station point about 2½ times the height of the building in front of the picture plane – being any closer than 2½ times distorts the drawing too much.

   The observer has a **cone of vision** of 90° at the most. He cannot see outside this area. The edges of the cone of vision go through the picture plane at the LVP and the RVP. Draw radiating lines from the SP to cut the picture plane at 1, 2, 3, etc. Draw in the verticals. These are the corner positions of the perspective drawing.

3*a* and 3*b*. Choose an eye-line. This may be high (3*a*) or low (3*b*). Each set of parallel lines meets at a vanishing point (VP) on the eye-line. Project down from the LVP and the RVP in (2) to the same positions in (3*a*) and (3*b*).

Draw the front edge of the house to the correct height Aa. Project to the vanishing points and line in the walls. Draw the height of the roof ac **at the front corner edge.** Draw from c to the LVP. Line 2 crosses this line at d. This is the height of the ridge. Draw from d to the RVP to cut 5 at e. Draw the roof.

Finally extend ad and extend fe to cut it at the **roof vanishing point** which we will use later.

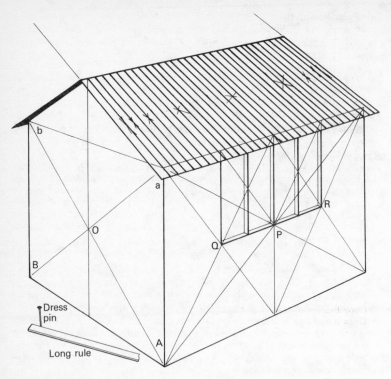

Fig. 9.7. Method of finding centres by repeated division

An Enlarged View of the Building Showing the Method of Finding Centres by Repeated Division

The method of finding the centre of 'perspective rectangle' is shown in Fig. 9.7. Draw the diagonals of rectangle AabB. Then O is the **visual centre** of the rectangle. The long side of the rectangle has been divided at P and again at Q and at R. The roof has been divided repeatedly at 1, 2, etc. Lines have then been drawn through these centre points from the roof vanishing point. The RVP is a long way outside the drawing so an extra board and a long rule had to be used.

Figs 9.8 and 9.9 show the effect of changing the station point. Fig. 9.9 is

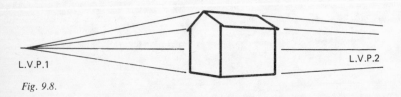

L.V.P.1   L.V.P.2

Fig. 9.8.

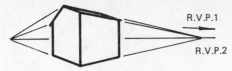

R.V.P.1

R.V.P.2

Fig. 9.9.

distorted because the station point is too close to the building.

Questions on Perspective

1 Find a famous picture which is painted in perspective and try to see how it was constructed. Try to work out where the artist would have been standing to draw it. Leonardo da Vinci worked out an elaborate formula for this but you should do it by common sense. There is no exact answer to this question. It is a question you will ask yourself all your life when you look at pictures. **Remember that some pictures are not in perspective at all.**
2 Copy Fig. 9.3 and, if you like, Fig. 9.4. You can vary the furniture in the kitchen but do not change the window.
3 Make a drawing of a long corridor with one vanishing point at the far end. Give it a square tiled floor drawn like Fig. 9.2, but long and narrow. Shade the squares in black and white.
4 Copy Fig. 9.6 on a large sheet of paper. Choose your own eye-level.
5 Draw a house of your own design using the same method. Then put another one beside it. Then a couple more of a different type. You can build up a street design if you like. Remember that the roads obey the same rules as the houses. They will have their own vanishing points. You can even work out the correct heights of people near to you and far off.

More difficult questions

6 Draw Fig. 9.7 and divide the sides and roof.
7 Draw a block of modern flats and draw the lines of windows correctly by repeatedly dividing. Remember that all measurements must be made at the front edges.

Pictorial drawing continues in Chapter 11 (page 195).

10
Solid Geometry IV

Continued from Chapter 8 (page 162)

Solids of Revolution

A series of solid figures produced by turning a flat shape round an axis, is shown in Fig. 10.1. Any solid figure which is symmetrical about an axis is a **solid of revolution**. One such example is a wooden bowl turned on a lathe.

Ibn-Tulin Mosque AD 876–879

Fig. 10.2 shows the Ibn-Tulin Mosque. The simple square building has

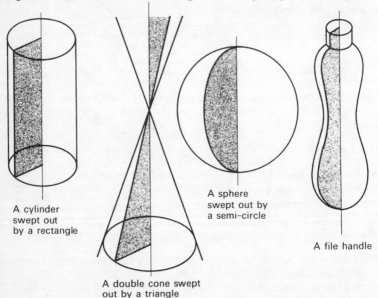

A cylinder swept out by a rectangle

A double cone swept out by a triangle

A sphere swept out by a semi-circle

A file handle

Fig. 10.1. Solids of revolution

Fig. 10.2. Ibn-Tulin Mosque AD 876–879

tall, pointed arches at ground level. Above them are large windows. The corners of the building are reduced in two steps to an octagon. On this rests the beautiful, severe dome. The dome is a solid of revolution. It was not made by turning of course, but it is given the name.

True Lengths and True Shapes

In Fig. 8.7 we saw that a triangle appears to change its shape as it turns. We can see a line as a true length *only* when it is at right angles to us. Therefore, we can see a line in true length by:

1 Turning the line so that it is at right angles to us.
2 Moving ourselves so that we are at right angles to the line.

Visual Aids

Visual aids like those shown in Fig. 10.3 are useful for deciding when lines are seen in true length and when they are not. Flat shapes have rods fixed to them. At home, fasten a paper triangle to a pencil with sticky-tape. As the pencil is revolved, the lengths of the sides of the triangle change. We can see the true shape of the triangle only when all the sides are seen as true lengths.

When the rod is horizontal, a true shape can be seen in plan twice in every revolution.

When the rod is parallel to the vertical plane, a true shape can be seen twice in elevation.

When a rod is inclined to the HP *and* the VP, we can see a true shape *only* in an auxiliary view.

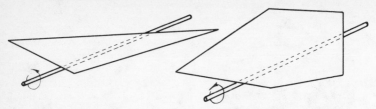

Fig. 10.3. Visual aids: Plane figures with rods fastened to them as axes

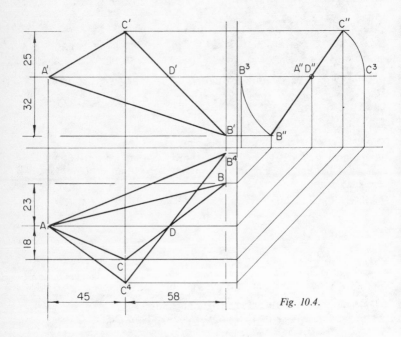

Fig. 10.4.

In Fig. 10.4, triangle ABC lies, as shown, inclined to the VP and to the HP. To see it in true shape, we must revolve it.

Method Draw AD parallel to the VP. Project to the elevation at A'D'. A'D' is also parallel to the HP. **This is unusual.** You will see why later. Project an end elevation. AD is seen as a point and BADC is a straight line. Revolve B"C" round A"D" so that it is horizontal at B³C³. Project the true shape AB⁴C⁴.

In Fig. 10.5 triangle ABC is shown inclined to the VP and the HP. To find the true shape use the following steps.

Method Draw line AD parallel to the VP. When we project up, A'D' is not parallel to the HP but slopes up. We need to revolve the triangle to find the true shape. To do this we must see A'D' as a point. We must take an

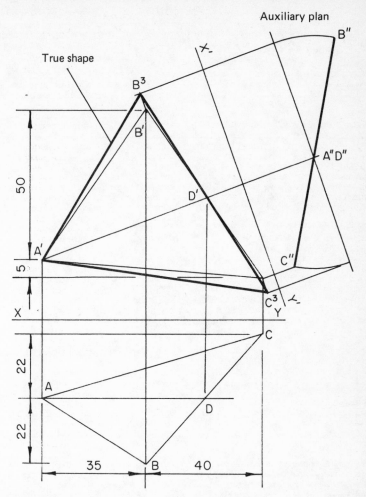

Fig. 10.5.

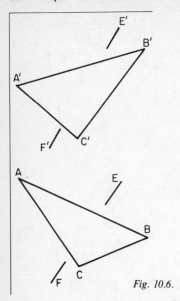

Fig. 10.6.

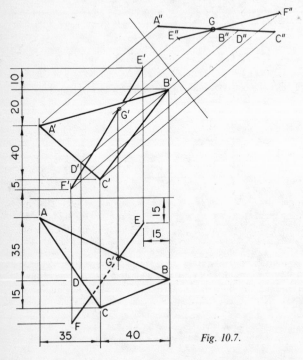

Fig. 10.7.

auxiliary plan looking along A'D'. Triangle ABC becomes the straight line B"C". Pivot round A"O". Project the true shape A'B³C³. This is, in effect, a rotation of the triangle about A'D' in elevation.

In Fig. 10.6, triangle ABC is pierced by a rod EF. Find the point where EF passes through the triangle.

Method Set out the triangle in plan and elevation as shown in Fig. 10.7. Draw line BD parallel to the VP. Project to B'D' in the elevation and project an auxiliary plan A"B"D"C" with E"F" passing through it at G. Project to G' and G". Line in. **Work out carefully which parts are hidden.**

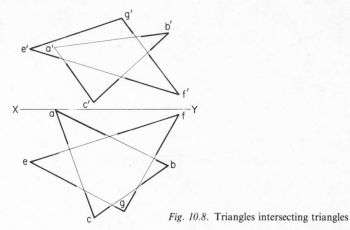

Fig. 10.8. Triangles intersecting triangles

Intersecting Solids

Triangles Intersecting Triangles

In Fig. 10.8 a triangle abc cuts through another triangle def. Only the tips of the triangles can be lined in as we do not know where the cuts will come. Fig. 10.9 shows the triangles redrawn. We now want to find where one triangle intersects the other.

Method Draw bd in the plan parallel to the VP. Project up to b'd'. Project an auxiliary plan looking along d'b'. Triangle abc is seen as a straight line which cuts e"f" at r and g"f" at s. Project to r' and s'. The triangles intersect (cut each other) along r's'. Project to the plan. Line in **carefully.**

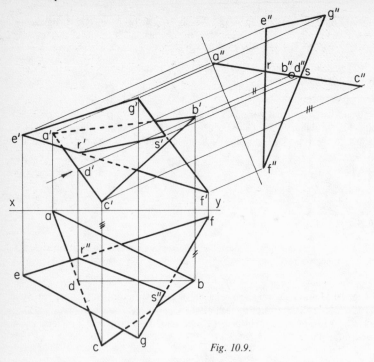

Fig. 10.9.

Fig. 10.10 reminds us that planes can be vertical, horizontal, inclined to the vertical plane and inclined to the horizontal plane. These planes (and one more) are the subject of this section.

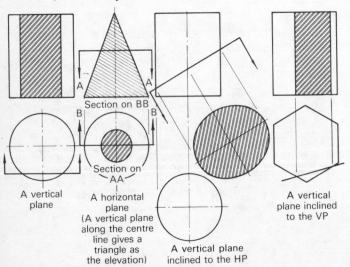

Fig. 10.10. A reminder of the cutting planes

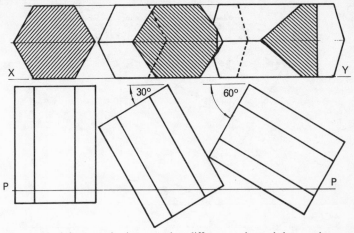

Fig. 10.11. A hexagonal prism turned to different angles and then cut by a vertical plane parallel to the VP

Fig. 10.11 shows a hexagonal prism in three different positions, cut by a vertical plane PP, parallel to the VP.

Fig. 10.12 shows a cylinder inclined at 30° to the HP and cut by a horizontal plane.

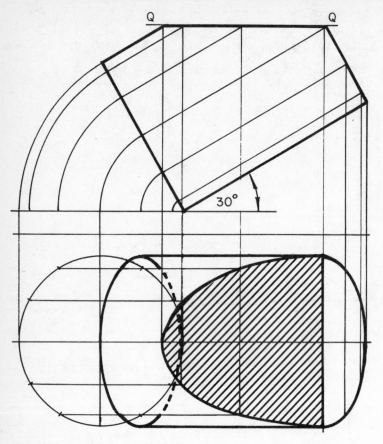

Fig. 10.12. A cylinder cut by the horizontal plane

Visual Aids

Two boards are hinged together to make a visual aid (as shown in Fig. 10.13). They have vertical and inclined grooves in them to hold thin plywood or mild steel sheets.

View A shows a vertical inclined plane. The grooves where the plane passes through the vertical and horizontal planes are labelled vth. These are the **vertical trace** and the **horizontal trace.**

View B shows a plane inclined to the horizontal.

View C shows a plane inclined to the vertical *and* the horizontal planes. It is an **oblique plane** and is the one referred to in brackets above.

   The small blocks illustrate how the planes cut solids. If the planes are

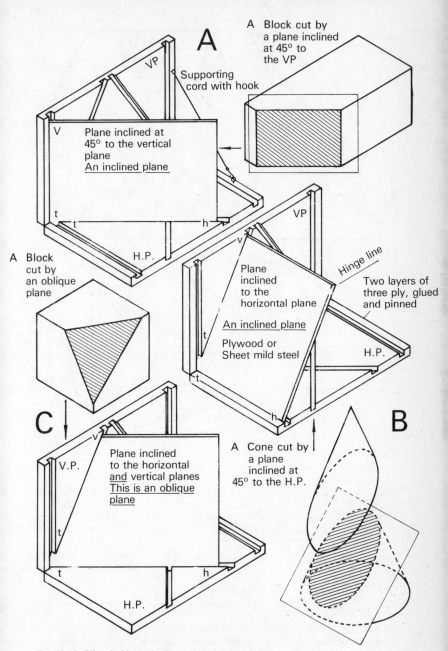

Fig. 10.13. Visual aids on inclined and oblique planes

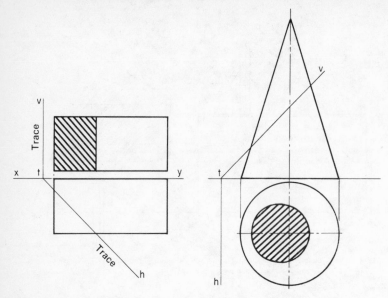

Fig. 10.14. A plane inclined
at 45° to the VP

Fig. 10.15. A plane inclined
at 45° to the HP

Fig. 10.16. A hexagonal prism cut by a plane inclined at 30° to the HP

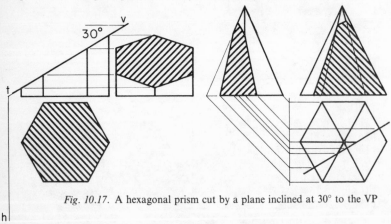

Fig. 10.17. A hexagonal prism cut by a plane inclined at 30° to the VP

made of mild steel and magnets are fitted to the blocks, the cut-off corners
will stay in place and make the demonstration more dramatic.

Fig. 10.14 shows a block cut by a plane inclined to the VP. The angle of
inclination to the VP **is shown in the HP**. The inclined plane cuts the VP
along the vertical trace vt, and the HP along the horizontal trace ht.

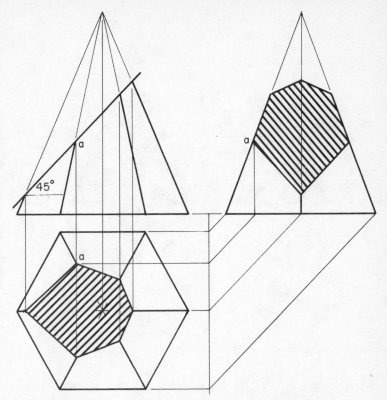

Fig. 10.18. A hexagonal pyramid cut by a plane inclined at 45° to the HP

Fing 10.15 shows a cone cut by a plane inclined at 45° to the HP. The traces are lines vth.

In Fig. 10.16 a hexagonal prism of 30 mm side is cut by a plane inclined at 30° to the HP. The traces are shown but normally we draw only the cut which is heavily lined in.

In Fig. 10.17 a right hexagonal pyramid of 50 mm side and 100 mm high, is cut by a vertical plane inclined at 30° to the VP.

In Fig. 10.18 a hexagonal pyramid of 40 mm side and 90 mm high, is cut by a plane inclined at 45° to the HP.

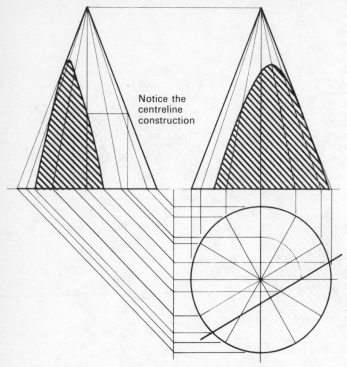

Notice the
centreline
construction

Fig. 10.19. A right cone cut by a plane inclined at 30° to the VP

Fig. 10.19 shows a right cone 80 mm diameter and 90 mm high, which is cut by a plane inclined at 30° to the VP and 10 mm from the centre line. In Fig. 10.20 a right cone is cut by a plane inclined at 30° to the HP.

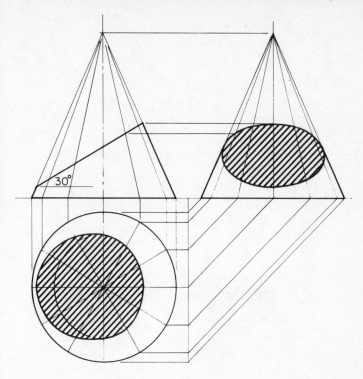

Fig. 10.20. A right cone cut by a plane inclined at 30° to the HP

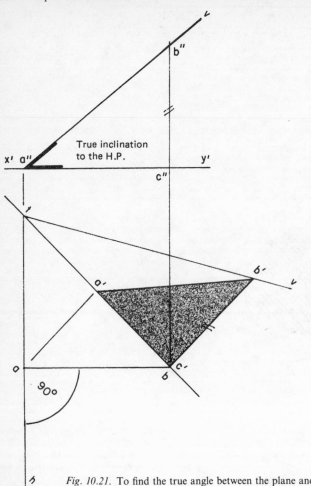

Fig. 10.21. To find the true angle between the plane and the HP

An Oblique Plane

In the visual aid (see Fig. 10.13 view *c*), *both* traces are inclined. The plane is inclined to the HP *and* the VP.

To Find the True Angle Between the Oblique Plane and the HP

(See Fig. 10.21.) Draw a line at right angles to the ht as far as the xy line. This is the plan view of triangle a′b′c′. This triangle fits between the plane and the HP at right angles to the horizontal trace. Therefore angle bac is the true angle between the plane and the HP.

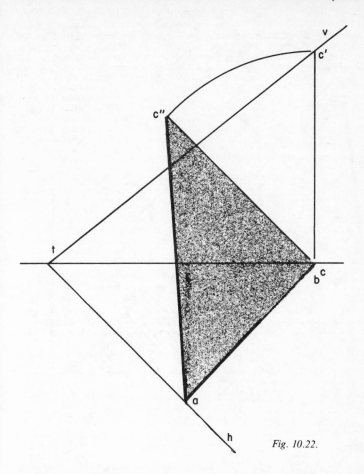

Fig. 10.22.

Method Look along the ht so that it is seen as a point a″. Draw the x′y′ line at right angles to ht. Transfer the height c′b′ to c″b″ and draw the edge view of the vertical trace vt. The angle b″a″c″ is the true inclination of the plane to the HP.

An **alternative method** is shown in Fig. 10.22. Once you understand the previous method you can draw it this way, which is slightly quicker. Here the triangle, which is seen as a line abc in plan, has been laid flat.

In Fig. 10.23 a small rectangular prism 60 mm × 30 mm × 25 mm lies on the horizontal plane, 5 mm in front of xy line. The traces vth from Fig. 10.21 have been redrawn as shown. An auxiliary view along the horizontal trace shows that corner p has been cut off. Project back to the plan at qq and up to the elevation. Project the front and back of the plan along to the ht at r and s, and then up to the elevation. Measure the angle between the oblique plane and the HP.

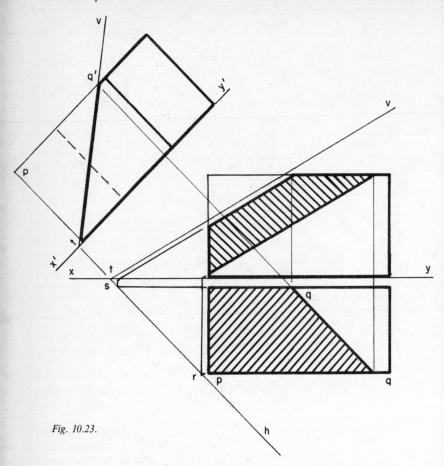

Fig. 10.23.

A triangle a'b'c' is shown (see Fig. 10.24) in front elevation resting in an oblique plane. The measurements are given in Figs. 10.24 and 10.25. Draw the plan.

Solution Take an auxiliary elevation along ht as shown in Fig. 10.25. Triangle a"b"c" is seen as a straight line. Measure the heights of a, b and c in the auxiliary view. Project back and draw the plan. Measure the correct angle between the plane and the HP (b"x'y').

Questions

All the examples in this section should be drawn. Where measurements are not given, make them at least double the book size and four times bigger for some. It is important to use your imagination. Get some small blocks,

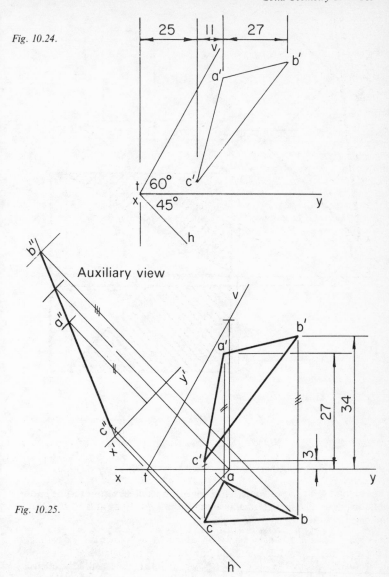

Fig. 10.24.

Auxiliary view

Fig. 10.25.

even if they are not the correct shapes, and hold pieces of card as if to cut them. Try to visualise in your mind what is happening as the planes slice the blocks. With a lot of practice, you will learn to see the solution in your mind before you start drawing.

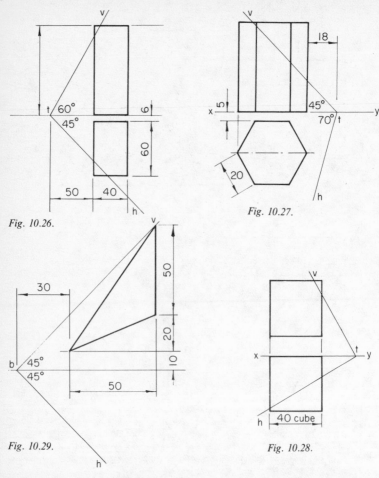

Fig. 10.26.

Fig. 10.27.

Fig. 10.29.

Fig. 10.28.

Figs 10.26–10.28 are typical examination questions. In each case, project an auxiliary view and then draw the front elevation and plan of the piece of the block below the oblique plane.

Fig. 10.29 completes the plan by drawing an auxiliary view and projecting back.

This is advanced work which does not normally come into many 'O' CSE level syllabuses, but is very interesting to tackle if you have time.

Double Auxiliary Views

In Fig. 10.30 an auxiliary plan is projected from an elevation. An auxiliary elevation is projected from the auxiliary plan. The heights above x^2y^2 are equal to the heights above x^1y^1.

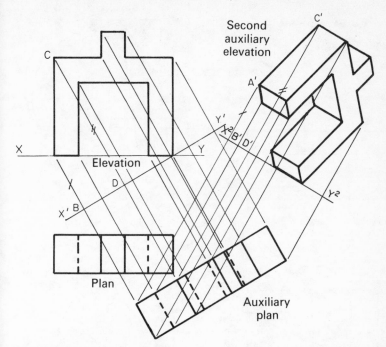

Fig. 10.30. Double auxiliary view

In Fig. 10.31 an auxiliary elevation has been projected from A. An auxiliary elevation has been projected from B and an auxiliary plan C from the auxiliary elevation on B.

Fig. 10.31. Double auxiliary view

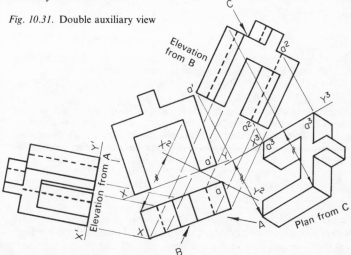

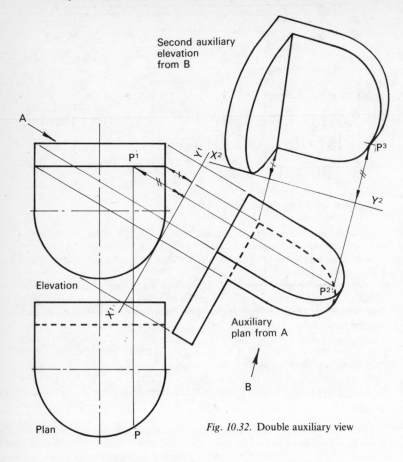

Second auxiliary
elevation
from B

Elevation

Auxiliary
plan from A

B

Plan

Fig. 10.32. Double auxiliary view

In Fig. 10.32 auxiliary elevation B has been projected from the auxiliary plan A. The projection of one point on the curve p is shown.

Questions
All of the questions in this chapter should be drawn double size or larger. Most of the questions depend on understanding when lines are seen as true lengths and when they are not. **It is most important for students to learn this material carefully by actually turning small objects around and working out the questions by watching the lengths change.**

Continues in Chapter 12 (page 208).

11
Pictorial Drawing IV – Axonometric or Planometric Drawing

Continued from Chapter 9 (page 173).

Architects like axonometric projection because it is a picture drawing which is easily understood by everyone, but which has a true plan and can have true vertical heights. In the one drawing they have a picture drawing and an accurate measured drawing.

Axonometric means 'measured round an axis (axon)'. Imagine a cube shaped box lying on the ground with one corner pointing towards you. The bottom is the plan view and will be drawn correctly. The lid is then pushed away from you until the diagonal line from the back bottom corner to the front top corner becomes vertical. This line is the axis (axon) round which everything is drawn. It is something like one half of one of those tool boxes with top trays. The trays move away sideways but stay level.

Some of the drawings in this chapter look very elaborate but they are actually quite simple when you understand them.

Rules

1 Draw the plan with one corner towards you but with its true shape.
2 Erect perpendiculars and measure the heights. All vertical lines stay vertical.
3 Join the tops.

In Fig. 11.1 the plan and elevation of a simplified house is shown. Then

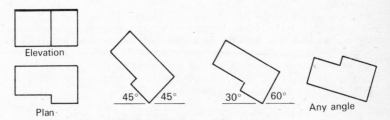

Fig. 11.1. Axonometric projection

Note: The plan is normally tilted at 45°- or 30°/60° but can be placed at any angle

Fig. 11.2. An axonometric housing site layout

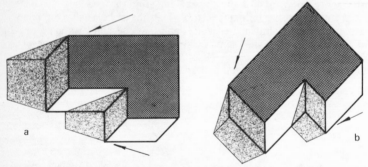

Fig. 11.3. Oblique and axonometric projection compared
       (*a*) Oblique projection with shadows (sometimes called planometric projection)
       (*b*) Axonometric projection with shadows

the plan has been tilted to different angles ready for drawing the vertical walls.

A simplified housing layout (see Fig. 11.2) shows houses set out in different directions. This is bad architectural planning because some kitchens and living rooms will be badly placed for the sun.

Oblique and Axonometric Projection Compared

In oblique projection the plan is out of shape. In axonometric projection

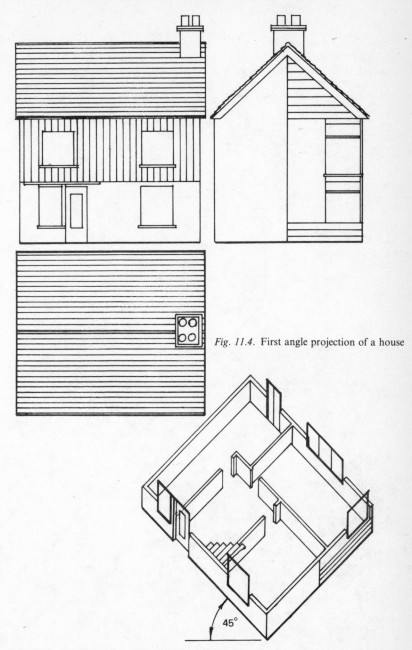

Fig. 11.4. First angle projection of a house

Fig. 11.5. Axonometric projection of the ground floor

plans are always accurate. Shadows of simple blocks are easy to apply (see Fig. 11.3).

Fig. 11.4 shows the plan and elevations of a house in first angle projection. Fig. 11.5 shows the plan turned at 45° – nothing else has changed in the plan. Vertical lines have stayed vertical. The walls have been cut off about 1½ metres high. This is a favourite method of showing the ground floor of a house.

Fig. 11.6. Axonometric outline of a house. Sometimes called a glass drawing

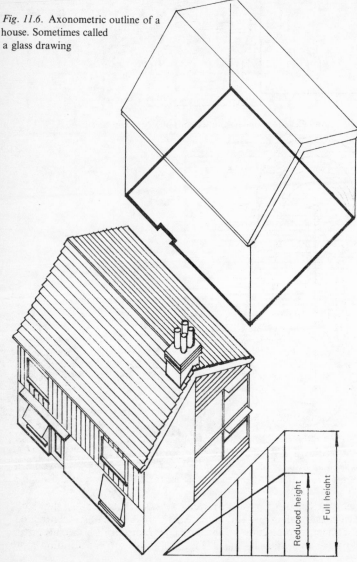

Fig. 11.7. A reduced height scale

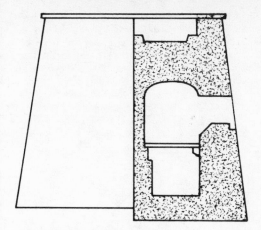

Fig. 11.8. Front elevation (one half in section) of a Martello tower

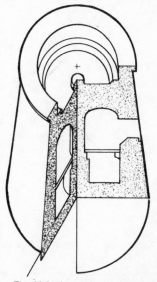

Fig. 11.9. Axonometric projection of a Martello tower

In Fig. 11.6 the same house, with the plan lined in heavily and a skeleton above it, is drawn at a **reduced height**. Fig. 11.7 shows the house complete. The drawings often look more natural when the wall heights have been reduced. Here they have been shortened in the scale shown at the side of Fig. 11.7.

During the Napoleonic wars, Martello towers were built along the south and east coasts of England and in places in Canada, South Africa, Ireland and elsewhere, as protection against invasion. James Joyce's famous novel *Ulysses* starts in a Martello tower. Fig. 11.8 shows a front elevation (one half in section) of a Martello tower. An axonometric projection of a Martello tower is shown in Fig. 11.9. They were, in fact, slightly oval but this one has been drawn circular. The centres of the different levels of circles lie one above each other on the centre line. This is a very quick method of drawing circular objects. This drawing shows clearly that axonometric projection is really oblique projection with the plan in true shape.

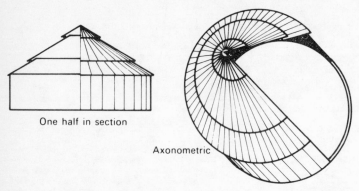

One half in section

Axonometric

Fig. 11.10. The Round House, Chalk Farm, London

Note: This was built as a house for turning railway engines for their return journeys from London

A. Choissy was an architect who taught for years at the Beaux Arts in Paris and wrote a number of books on architecture. Fig. 11.13 is a re-drawing, using modern architectural tints, of his drawing of the temple at Athos, in Greece. Here, Choissy drew in axonometric projection but looked upwards, through the floor. Thus he was able to show inside the domes. This is a very useful method, sometimes used to show foundations.

Fig. 11.11. An axonometric projection of a stucco house on the Regent's Canal, London

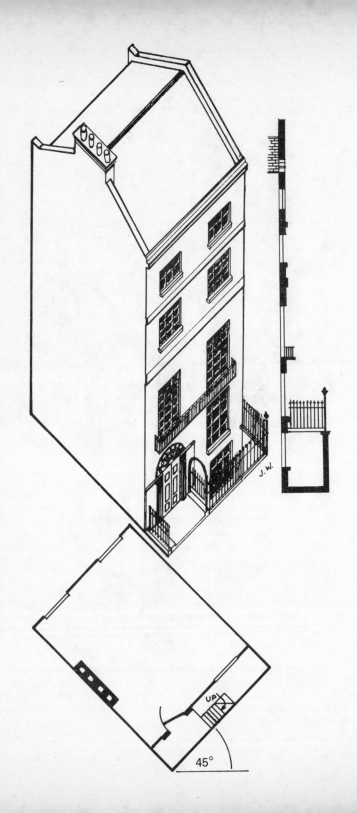

J.W.

45°

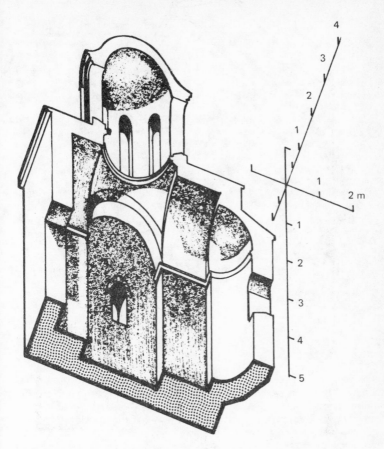

Fig. 11.13. An axonometric projection of the temple at Athos from below (A. Choissy)

The Forma Urbis Severiana

The Forma Urbis Severiana was a plan of Rome, carved in marble, made between AD 203 and 211, and placed on the wall of Vespian's Forum as a sort of legal document describing the buildings. It is now in fragments. More than a thousand pieces are known. Some have been fitted together to form larger sections but few have been identified and related to particular places in Rome.

We can however, work out some of the types of buildings from the

Fig. 11.12. An eighteenth-century house with a basement and a section of the front wall

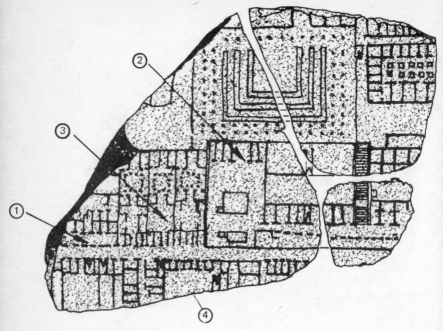

Fig. 11.14. Fragments of the Forma Urbis Severiana

plans engraved on the marble. Pages 203–5 are an attempt to show what the engravings may mean. The ideas for these drawings came from *The Golden House of Nero* by A. Boethius, Michigan University Press, 1960. I found the book fascinating, but I must stress that the drawings are mine and in no way authoritative.

Some pieces of the Forma Urbis are shown in Fig. 11.14.

Fig.s 11.15 shows possible interpretations of the row of shops (tabernae) seen as arrow 1 on Fig. 11.14 of the Forma Urbis, with a covered walk at the front to give shade at midday.

Fig. 11.16 gives a possible meaning of 2 on the engraving, where there is no covered way in front and a gap. This might be a wider shop but it would be a very wide opening for a shopkeeper to have been able to afford in those days.

The Forma Urbis drawings are included in the hope that students will feel encouraged to try to reconstruct buildings from their remains. A skilled archaeologist, with only a few post holes, can reconstruct a building. Petrie, the Egyptologist, found a ruined temple site. The stones had been taken away by subsequent generations. All he had to work from were the red marking-out lines, set out on the ground by the original builders. From them he managed to draw an acceptable reconstruction of the temple.

Your local Archaeological Society may have a site worth reconstructing

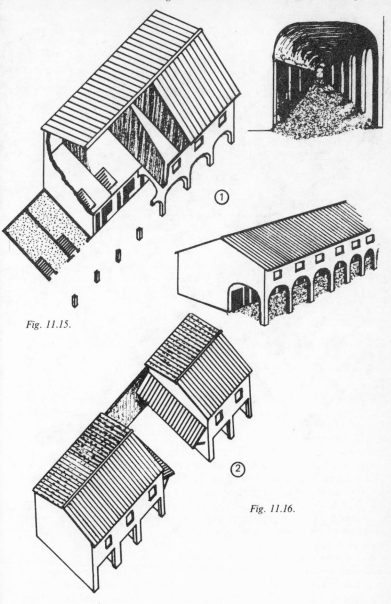

Fig. 11.15.

Fig. 11.16.

in axonometric projection. The local cricket pavilion may be in ruins, but it might make a good drawing.

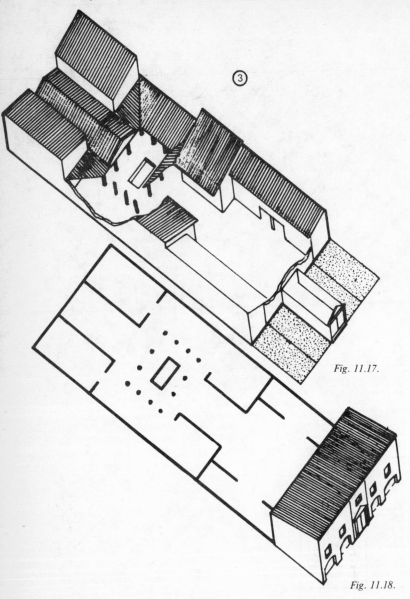

Fig. 11.17.

Fig. 11.18.

Fig. 11.17 gives a possible drawing of the houses in 3. They had a central entrance from the street to a courtyard. The shops on either side (see Fig. 11.18) were separate, lock-up shops.

Fig. 11.19 shows possible interpretations of the large appartment block in 4. This seems like the existing ruins in Ostia.

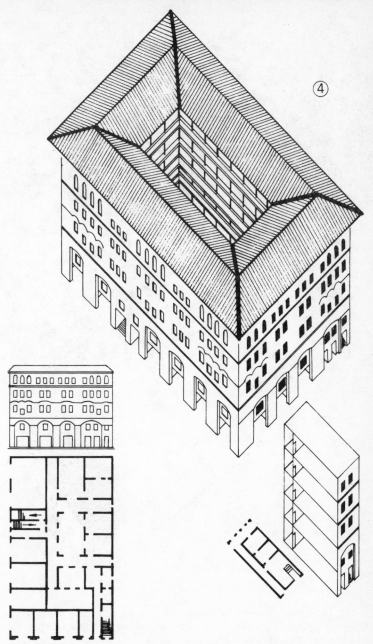

Fig. 11.19.

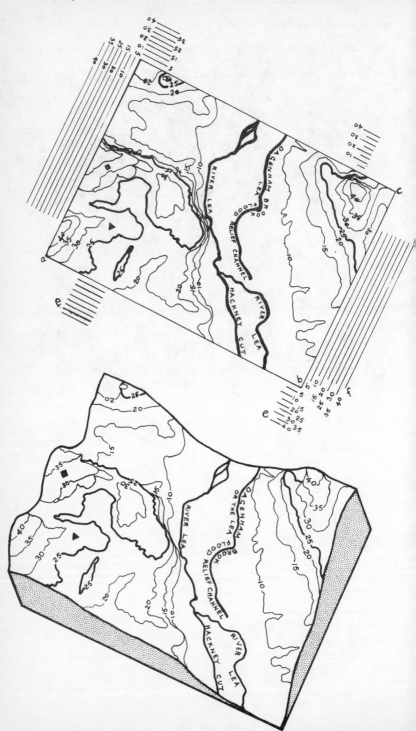

Axonometric Block Maps

A three-dimensional map is easy to make, especially if the contour map with spacing lines is supplied.

Fig. 11.20 shows a contour map of the River Lea Valley, drawn to a small scale. It was actually produced by repeated reduction on a photocopier. It has been tilted at 45° and lines have been drawn parallel to the edges and numbered as shown.

Fig. 11.21 is a block map showing the rise and fall of the land. It has been smoothed out but should be thought of as a series of layers. Each layer is a contour height.

Method Place a piece of tracing paper over Fig. 11.20 and trace the bottom lines of the map ab and bc. Move them down to the 40 m lines ee, ff. Trace the 40 m contour lines. Move lines ab, bc up to the 35 m lines and draw the 35 m contour. Do not draw behind the 40 m contour lines. Continue to move the tracing paper up and draw the contours in turn. When all the contours have been drawn, join the ends of the the contour lines smoothly to give the irregular outline. Draw the bottom edges of the block at 45° and the vertical ends. Shade the vertical sides. Number the contours to read from low to high. Darken in every fifth contour.

The flat contour map Fig. 11.20, can be viewed from any corner to give different block maps. Thus a class can produce four views of the same area.

Axonometric projection is not in most examination syllabuses but it is a very useful projection, essential for architects and suitable for many craft design and technology drawings. It is used widely in advertising and graphics. Soon the examiners will fall in line.

Fig. 11.20. A contour map of the River Lea Valley with height lines for tracing

The contour map with height lines for tracing. The map can be viewed from any corner. Between them, a class can produce all the views. The plan was made by lining in contours heavily and repeatedly reducing the map on a photocopier. Finally a good copy was made.

Fig. 11.21. An axonometric block map of the River Lea Valley

Note: Such drawings can be used as the basis for solid models in plywood, plaster, etc.

12
Solid Geometry V

Continued from Chapter 10 (page 164)

A section is a cut. Fig. 12.1 shows a variety of cutting planes and the shapes they produce.

1 A cylinder cut parallel to the centre line produces a rectangle.
2 A cone cut parallel to the base produces a circle.
3 A cylinder cut obliquely gives an ellipse.
4 A prism cut parallel to the centre line gives a rectangle.

Numbers 3 and 4 are cut by inclined planes.

The Canal Engineers used inclined planes (slopes) to drag their barges from one canal to another canal at a different level. A huge tub was sunk in one canal, the boat was moved over it and the tub was raised with the boat floating in it. Then the tub and boat were pulled up one slope and lowered into the other canal. A second tub, also full of water, acted as a counterweight. Fig. 12.2 shows a typical inclined plane.

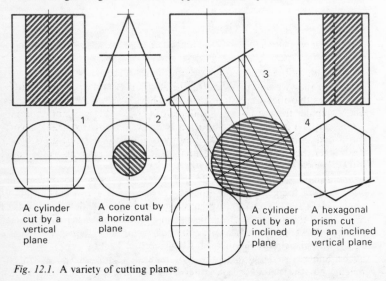

1   A cylinder cut by a vertical plane

2   A cone cut by a horizontal plane

3   A cylinder cut by an inclined plane

4   A hexagonal prism cut by an inclined vertical plane

Fig. 12.1. A variety of cutting planes

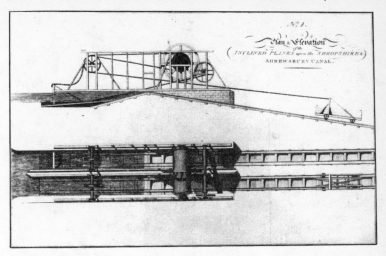

Fig. 12.2. An example of an inclined plane

The title of the engraving reads, 'No. 1. Plan and Elevation of the INCLINED
PLANES upon the SHROPSHIRE & SHREWSBURY CANAL'

In this chapter we are concerned with finding what happens when one
solid is cut by another but, before we start, it is worth telling the story of
Flatland. This is a famous book first published in 1884.

The Story of 'Flatland: A Romance of Many Dimensions', by A Square

(*Edwin Abbott Abbott*) Pub. 1884, reprinted 1926, 1932, Basil Blackwell

In Flatland, all the inhabitants were flat figures. The more sides they had,
the higher they were in the society. The aristocracy and priesthood had so
many sides that they were almost circles. The servants were mere triangles,
while irregular triangles were considered of very low class, if not criminals.

One day a Flatland mathematician, who was a Square, heard a voice
and found a circle appear beside him mysteriously. The circle grew bigger
and bigger, became smaller, and finally disappeared, but his voice could
still be heard.

The Square was bewildered. He had never seen a circle, or indeed any
other figure, change its size in this way. The voice tried to explain about
height. He was a Sphere and was **passing through** Flatland. All the Square
could see of the Sphere was a circle getting bigger and smaller. Circles near
the equator were large: circles near the poles were small. This was why the
Square saw the circles changing in size. (See Fig. 12.3.)

The Sphere tried to make the Square understand what is meant by
'height' but the Square could not do so. At last, in anger, the Sphere

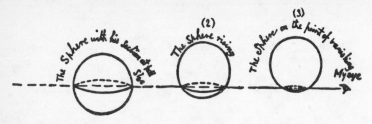

Fig. 12.3. The Sphere passing through Flatland. From the original illustration in *The Story of Flatland* by Edwin Abbott Abbott

By permission of Basil Blackwell Publisher

dragged the Square up, out of Flatland. Suddenly, the Square could see shapes, circles, polygons, squares, triangles, laid out before him. At last he had some idea of height.

The Sphere claimed that there were other lands with other dimensions. There could be four, five, twenty dimensions, if only one could imagine them. He took the Square to Lineland, where there was only length and people lived in a perpetual traffic jam, singing part-songs, as this was the only method of communication.

Next they went to Pointland where there were no dimensions and there was room for only one person. He was the King, and was so proud and self-satisfied, that when he heard the Sphere and Square talking, he thought he was a ventriloquist. (I have met people like that.)

Finally, the Sphere returned the Square to Flatland. The Square tried to convince Flatland that there was such a thing as height and was promptly declared insane.

How does *Flatland* help us to understand intersection problems? I think it helps our imaginations.

In Flatland, there was length and breadth. Nobody could look up or down. There could be a dozen parallel Flatlands, one above the other, knowing nothing about each other. When we solve an Intersection problem, we pass the solid through a series of parallel Flatlands and look at each section separately.

Flatland is well worth reading. It has some unpleasant, anti-feminist remarks which must be ignored but the geometrical ideas are amusing.

Two dimensional worlds have been taken further by Alexander Dewdney in *Planiverse: Computer Contact with a Two Dimensional World*, Pan, 1984. The creatures here are quite extraordinary, with nerve fibres 'crossing over' each other by computer gates, and two arms on each side of the body for gripping as they cannot bring an arm across the body from the other side.

Intersections of Solids

A Horizontal Cylinder Intersecting a Cone

Consider Fig. 12.4. Imagine the cone and cylinder passing upwards through

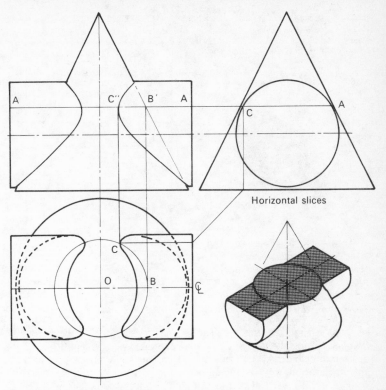

Fig. 12.4. A horizontal cylinder intersecting a cone

Flatland. The cone would be seen as a circle and the cylinder as a rectangle. At first there would be a series of circles growing bigger. Suddenly a line would appear. This would be the top centre-line of the cylinder. The line would become rectangles, which were first wider, then narrower, and finally diappearing, leaving circles. The circles would become larger and then disappear.

The circles and rectangles are easy to draw. If the cone and cylinder went through Flatland at an angle, the shapes would be much more difficult to draw. The cone would give ellipses and the cylinder pieces of ellipses. **Always make sure that you cut the figures so that the sections are easy to draw.**

The Method of constructing the lines where the cone and cylinder cut each other is shown in Fig. 12.4. (These curves are called **lines of intersection**.) AA is a horizontal section which cuts the cone at B′. Project down to B on the centre line of the plan. Then OB is the radius of the circle cut through the cone by AA. Draw the circle. AA cuts the end of the cylinder at C. Project to the plan to cut the circle at C. This is a point on the intersection in plan. Project up to the front elevation at C″. Draw other horizontal sections like AA to find further points of intersection and join up.

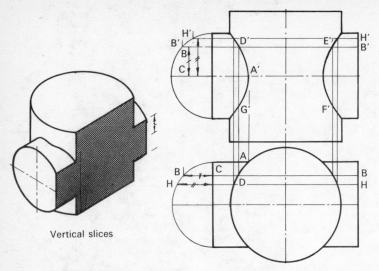

Vertical slices

Fig. 12.5. A vertical cylinder is intersected horizontally by a smaller cylinder

The Intersection of Two Cylinders

Method In Fig. 12.5 the plan shows the outside edge of the horizontal cylinder cutting the circle at A. This edge must be the centre line of the horizontal cylinder in the elevation. Project up to A′. Instead of drawing an end elevation, we have folded out semicircles in plan and elevation. BB is the edge of a **vertical** cutting plane. The width of the cut is twice BC. Measure BC from the centre line in the elevation. Draw B′B′. Project up from the plan to D′E′F′G′. Take other vertical sections HH, etc. and draw the lines of intersection.

The Intersection of a Cone and a Cylinder

This is the same problem as Fig. 12.4 so we could use horizontal section lines, **but this time we want to develop it**. A horizontal section will not be suitable for developing the cone. We need another way.

Method Divide the plan of the cone into twelve equal parts and number them (see Fig. 12.6). Project to both elevations and join to the apex. Line 0′10 cuts the cylinder at A″ and B″. Project across to A′ and B′, and down and around to A and B. Complete the lines of intersection by taking similar points.

Develop the cone and measure off *true lengths* measured along O″10 at g, f, d, etc.

Develop the cylinder by drawing the centre line. Set off lines C, E, A, etc. by stepping off the chords from the end elevation. Then measure the widths from the arrows in the plan.

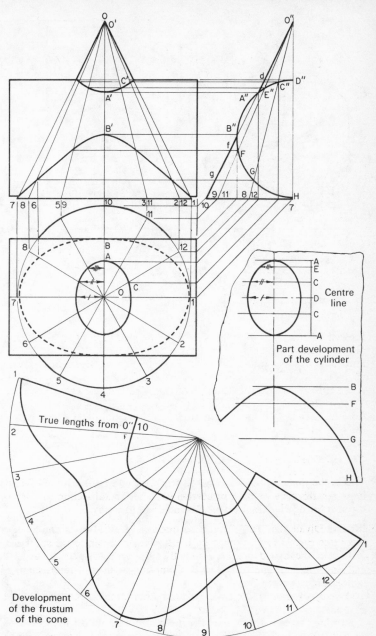

Fig. 12.6. The development of a cone intersecting a horizontal cylinder

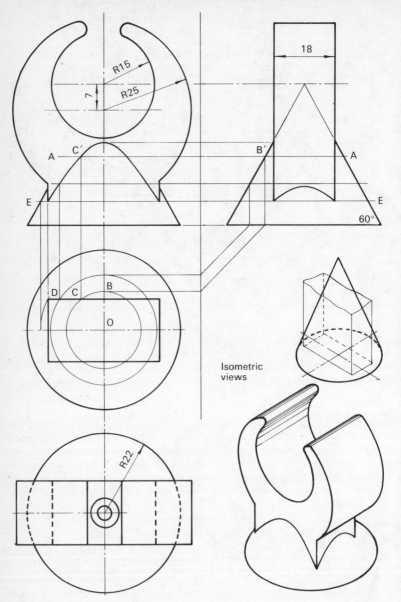

Isometric views

Fig. 12.7. A plastic domestic heating pipe clip

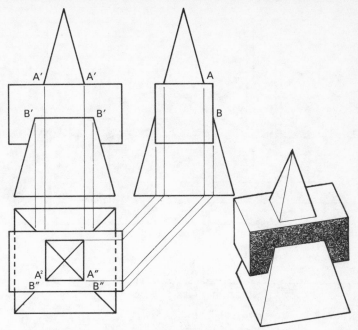

Fig. 12.8. A horizontal square prism intersecting a square pyramid

A Pipe Clip: The Intersection of a Cone and a Rectangular Prism

The pipe clip is shown in four views in Fig. 12.7. The upper plan shows just the cone and prism. Horizontal sections will be best as they give circles and rectangles in plan. Section AA cuts through the cone at B′ giving radius OB in plan. Draw this circle to cut the prism at C and project up to C′. Corner D gives radius OD in plan. Swing round to the centre line and project up to give section line EE and point D′. Complete the lines of intersection.

A Horizontal Square Prism Intersecting a Square Pyramid

Method Draw three views (see Fig. 12.8). We could think of cutting planes but it is easier to think of needles pushed through at each corner. Needles A and B are seen as points in the end elevation. Project round, thinking of each corner separately, and join up.

An Oblique Prism Cutting a Vertical Prism

(See Fig. 12.9.) Draw two views. Think of needles again. AA cuts the vertical prism in plan at B and C. Project up. Line in carefully.

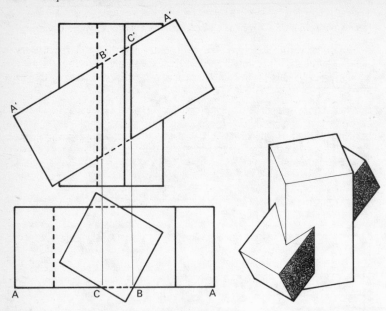

Fig. 12.9. An oblique prism cutting a vertical prism

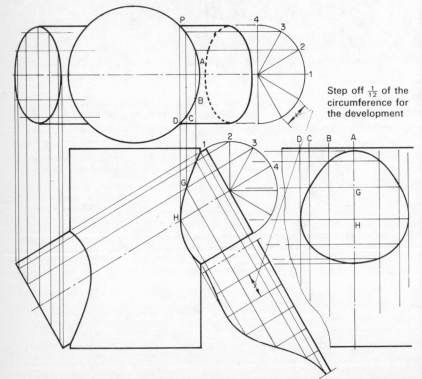

Step off $\frac{1}{12}$ of the circumference for the development

Fig. 12.10. The intersection of a vertical cylinder and a smaller oblique cylinder

The Intersection of a Vertical Cylinder and a Smaller Oblique Cylinder

We shall use vertical sections. We want to develop the two cut cylinders so we divide the smaller cylinder into twelve equal parts. We could draw any section lines but developments are easier with **equal divisions**; as they save time.

Method Draw semicircles on the ends of the small cylinder in plan and elevation (see Fig. 12.10). Divide both into twelve parts. Take vertical sections through the divisions 1, 2, 3, 4, etc. Project down from where they cut the large cylinder. Section 4 cuts the large circle at D and goes down to H. Complete the intersections.

Develop the small cylinder by rolling out at 90° and step off chords equal to $\frac{1}{12}$ of the circumference.

Develop the large cylinder by rolling out horizontally and stepping off the chords AB, BC, CD, etc.

Questions

Solve the problems in Fig. 12.11 with sizes at least four times those in the book. Collect small objects (like the pipe-clip) where one solid intersects another. Examine the lines of intersection and make sketches.

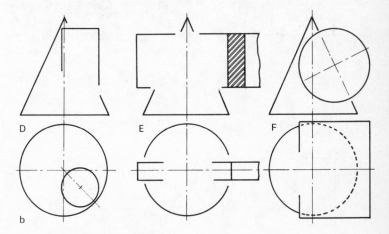

Fig. 12.11. Questions on intersections
   (*a*) Intersections of cylinders

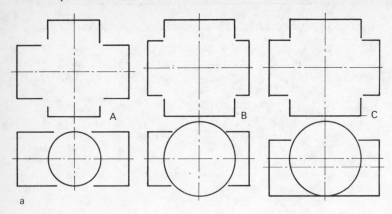

(*b*) Intersections of cones

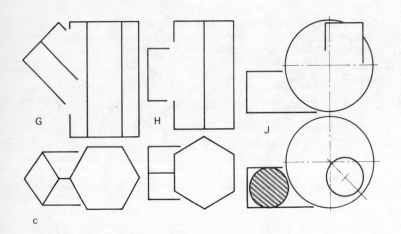

(*c*) Intersections of prisms

13
Loci or the Paths of Points

A locus (singular) is the path of a point. When a machine works, different parts move in different ways. (Locus is pronounced 'lokus', with a hard c.)

Fig 13.1 shows a diagram of a part of a machine. A is the fixed centre of a wheel. P is a point on the circumference. When the wheel turns, P marks out a circle. The locus, or path, of P is a circle. Lever CD pivots on C. The locus of D is the arc of a circle. As P revolves, it moves arm PC. The locus of C we shall find out later. As CD swings in its arc, it slides F along the groove. The locus of F is a straight line.

Every machine is a mass of loci (plural – soft c as in hiss) and they must be exact. If the loci interfere with each other, the machine will jam. The loci which follow are typical paths to be found in engineering, astronomy, biology and everywhere where things move regularly. It is true that a firefly marks out a path, but it is so irregular that the path has no name and is of no use to engineers because it cannot be repeated. A locus must be a regular, repeatable shape to be of any value.

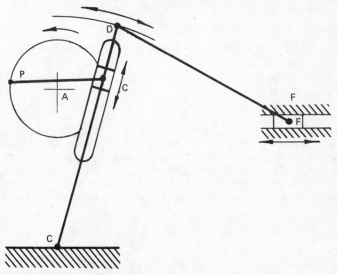

Fig. 13.1. A diagram showing the action of a machine

Fig. 13.2. Women flour millers portrayed by Sukenobu (1671–1751)
Courtesy of the British Library

Today, computers can be used to record random movements. A man
sprays one article with a paint spray. The computer records every movement and can then copy the process tirelessly for twenty-four hours a day.
We had no way of copying random movements like this before say 1960. It
is a complete revolution in engineering. However, random movements do
not concern us here, although there are hints of how they are recorded in
Chapter 19 which deals with computers.

'Women Flour Millers' by Sukenobu (1671–1751)

Simple machines show loci best as is illustrated by Fig. 13.2. The Japanese
women at the wood block, are milling flour in a traditional way. Three
women swing a trapeze. This pushes and pulls one end of a rigid bar, but
the other end can move only in a circular path. The fourth woman guides
the bar with one hand, so that it turns steadily in one direction, while her
other hand pours in grain.

Linear motion (motion in a line) has thus become **rotary** motion. A
motor car engine does the same. The expanding petrol mixture in the
cylinder forces out the piston. This drives the connecting rod, which turns
the crankshaft. One cylinder engines can get stuck at top, bottom or dead
centre. There is no Japanese lady inside the engine to keep it turning, so
we have several cylinders and one is always free to drive the crankshaft.

The Helix

A helix is the path of a point which moves round a cylinder and along it at
a fixed speed.

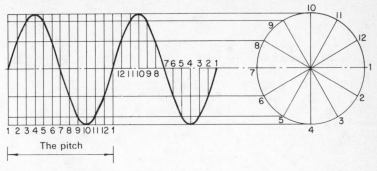

Fig. 13.3. A single helix

A Single Helix

This a screw thread.

Method Divide a circle into twelve equal parts and project them along the cylinder (see Fig. 13.3). The **pitch** of a thread is the distance from any point on a thread to the similar point on the next turn. Divide the pitch into twelve equal parts. Mark the positions 1, 2, etc. as the path moves round and along. Join up smoothly.

Visual Aid

An iron cylinder, such as a treacle tin, has a triangular rubber magnet wrapped round it to form a helix (see Fig. 13.4). The triangle is in fact a wedge. As a screw turns, it forces itself along as if it was rising up an inclined plane: a screw thread is a wedge shape wrapped round a cylinder.

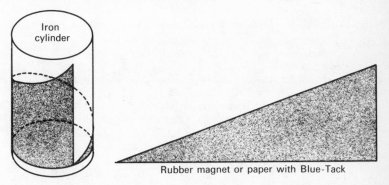

Fig. 13.4. Visual aid: a simple helix wound round a cylinder

A Double Helix or Two Start Thread

Two different helices (plural of helix, pron. 'heliseas') start from opposite sides of the cylinder (see Fig. 13.5). It is safest to draw them one at a time. This shape is now famous as the structure of the DNA molecule.

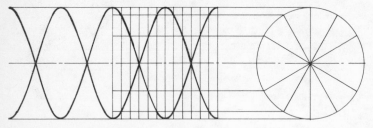

Fig. 13.5. A double helix or two start thread

A Helical Ribbon

(See Fig. 13.6.) Draw a single helix. Then mark the ribbon width beyond each point and draw the second helix. Line in carefully as some parts are hidden.

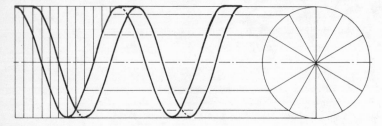

Fig. 13.6. A helical ribbon

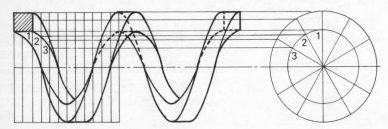

Fig. 13.7. A helical spring of square steel

A Helical Spring of Square Wire

Fig. 13.7 needs very careful drawing. Draw a helical ribbon. Then draw a second helical ribbon for the inside surface. Line in very carefully.

A Helical Spring of Round Wire

(See Fig. 13.8.) Draw the helix of the centre line of the wire. Draw circles round the helix of the centre line all along and join up the outsides.

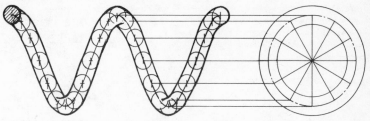

Fig. 13.8. A helical spring of round wire

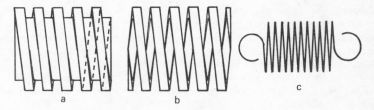

Fig. 13.9. Conventional thread and spring forms
        (*a*) A conventional square screw thread
        (*b*) A conventional compression spring
        (*c*) A conventional tension spring

Conventional Thread and Spring Forms

These are used to save drawing time (see Fig. 13A). They stand for the detailed shapes you have drawn. Make sure you know the difference between how a compression and a tension spring work.

The Archimedean Spiral

This is the path of a point which moves round a fixed point and outwards at a fixed rate. It is like the groove in a record disc.

Method Draw lines at 30° through the centre point 0 (see Fig. 13.10). Let OR be the distance travelled from the centre in one full turn. Divide OR into twelve equal parts. Draw arcs 1–1, 2–2, etc. as shown and join up the curve. The scroll plate of a three-jaw self-centring chuck is an Archimedean Spiral (Fig. 7.42).

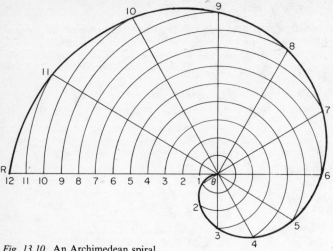

Fig. 13.10. An Archimedean spiral

Involutes

Involutes are the paths marked out by the end of an unwinding string which is pulled tight all the time. The visual aid (see Fig. 13.11) should make this clear. A circular block screwed to a board, holds a cord equal in length to the circumference of the block. A pencil is held in the looped end of the string. The cord is kept tight as it unwinds so that the pencil marks out an involute.

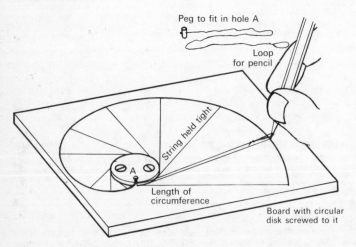

Fig. 13.11. Visual aid of an involute

Teachers will find that visual aids of geometry found in nature or in architecture (see Fig. 13.12) are always attractive and repay the effort of making them. The different forms of the helix are quite common: a helix is a spiral wound round a point or a line. The Archimedean spiral could be illustrated with a picture of an Egyptian shadoof (a device for raising water from a well). A lead screw of a lathe is an example of a square thread. A ram's horn grows in a spiral but gravity pulls it down as it grows so that the axis becomes a curve. An ammonite fossil illustrates an animal adding segments,

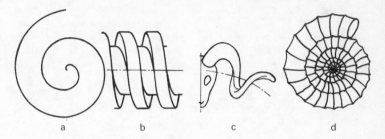

Fig. 13.12. Visual aid: different forms of the helix
  (a) An Archimedean spiral
  (b) A screw thread
  (c) A ram's horn
  (d) An ammonite

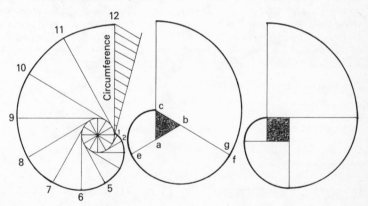

Fig. 13.13. Involute of a circle   Fig. 13.15. Involute of a square
Fig. 13.14. Involute of a triangle

each bigger than the one before. *Growth and Form*, by D'Arcy Thompson, has many examples of geometry in Nature. The spiral moves away from the centre at an increasing rate to give a snail shell, a horn shaped musical instrument, or a volute on an Ionic capital, as in the façade of the British Museum.

Method of Drawing an Involute of a Circle

(See Fig. 13.13.) Draw the circle and a tangent. Mark the length of the circumference along the tangent and divide it into twelve equal parts. Divide the circle into twelve equal parts and draw tangents. Draw arcs from 1, 2, 3, etc to cut the tangents as shown and join up the curve free-hand. This curve cannot be drawn with a compass.

Involutes of a Triangle and a Square

These *can* be drawn with a compass.

Method Fig. 13.14 shows how to construct the involute of a triangle. Extend sides ba, cb and ac. With centre a, swing ac round to e. With centre b, swing be round to g. Repeat for cg round c. At each corner the radius becomes bigger.

Construct the involute of a square similarly (see Fig. 13.15). Involutes of other figures can be constructed in the same manner.

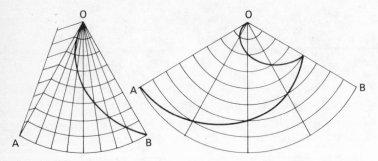

Fig. 13.16. A pendulum   *Fig. 13.17.* Another pendulum

Pendulums

In Fig. 13.16 a pendulum OA swings on pivot O to OB. As it swings a weight falls at a regular rate from O to B. Draw the path of the point.

Method Divide OA into any convenient number of equal parts. Divide the angle AOB into **the same** number of parts. Plot the path of the weight as it falls and swings.

Fig. 13.17 shows another pendulum. OA swings from A to B and back to A during the time the weight takes to fall from O to A.

Method Divide OA into a number of equal parts. Divide angle AOB into **half this number** of equal parts. Plot the fall.

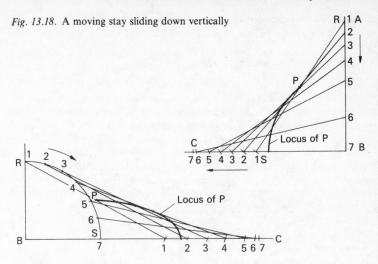

Fig. 13.18. A moving stay sliding down vertically

Fig. 13.19. A moving stay sliding round a quadrant

Simple Mechanical Examples

A Sliding Stay

A straight stay RS crosses a corner ABC (see Fig. 13.18). As R slides down towards B, S slides towards C. Set out any positions 1, 2, 3, etc. for R along AB. Mark off the length RS from each point on to BC with a *compass*. Draw the different positions of RS. P is the centre point of RS. Mark P for each position of the stay and join up these points to give the locus of P.

Fig. 13.19 shows another sliding stay. RS slides so that R follows the quadrant and S slides along BC. Draw the positions of RS and the locus of the centre point as shown.

Locus of the End of a Pivoted Arm

In Fig. 13.20 an arm PQ slides through a pivot R. P is driven by a wheel. Plot the locus of Q. This might be important in, for example, designing a safety guard for a machine with a moving arm.

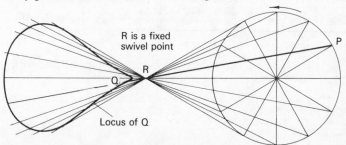

Fig. 13.20. Locus of the end of a pivoted arm

Method Take a series of points on the circle. Draw from each through R and mark the length PQ on each. Join to form the locus.

The Locus of a Point on a Connecting Rod

This time (see Fig. 13.21) the arm moves so that T travels backwards and forwards in a straight guide (not shown). Divide the circle and draw the position of ST for each 30° of movement. Mark P in each position and join up to form the locus of P. In which direction will P travel?

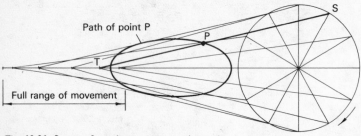

Fig. 13.21. Locus of a point on a connecting rod

The Cycloid

A cycloid is the path of a point on the circumference of a wheel, as the wheel makes one complete turn on a straight line. A nail in a cycle tyre probably has time to mark out one complete cycloid before the tyre bursts.

Method (See Fig. 13.22.) Draw the circle, divide into twelve equal parts and number them. Draw a tangent and mark out on it a length equal to the length of the circumference. Divide this length into twelve equal parts and number them. Draw lines through the points 2–12 on the circle parallel to the base tangent. Draw vertical lines from the points on the circumference up to the **centre horizontal line.** Do not go above.

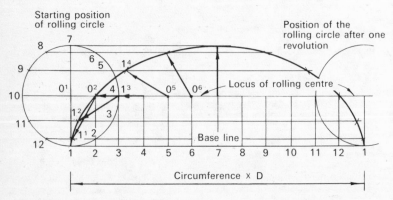

Fig. 13.22. The cycloid

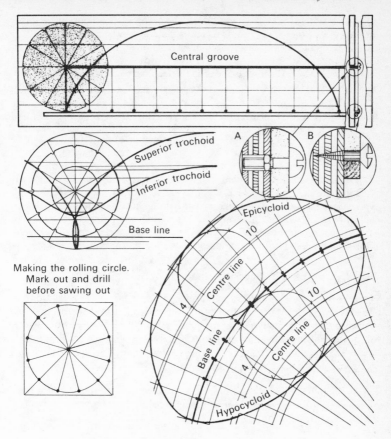

Central groove

Superior trochoid

Inferior trochoid

Base line

A   B

Epicycloid

10

Centre line

4

10

Base line

Centre line

4

Hypocycloid

Making the rolling circle.
Mark out and drill
before sawing out

Fig. 13.23. Visual aid: a board showing the formation of a cycloid
Fig. 13.23(a). Making one rolling circle
Fig. 13.24. Visual aid: showing the formation of epicycloids and hypocycloids
Fig. 13.25. The superior and inferior trochoid

Now imagine the circle rolling along the base line. When, for example, point 5 of the circle is on the base line, the centre will be at O^5. Point 1 will have reached the horizontal line 5–9. The arrowed radii show the progress of point 1 (the nail mentioned above) as it marks out the cycloid.

Visual Aid on the Cycloid

The cycloid demonstration board (see Fig. 16.23) consists of a rolling circle in tinted plastic which moves along a row of thirteen coach headed screws and a ledge below them. The circle, which is engraved into twelve segments, has semicircular slots in the circumference. These locate on the screws. One radius is heavily marked. The circle can be stopped at any point to show the position of the heavy radius which is generating the cycloid.

Enlarged section A shows the rolling circle with a nut and bolt through it which runs in the central groove of the board.

Enlarged section B shows one of the thirteen base screws with the rolling circle resting on it.

The epicycloid and hypocycloid

Fig. 13.24 shows how a point on the circumference of a circle rolling on a **straight line** produces a **cycloid.** A point on the circumference of a circle rolling on the **outside** of another circle produces an **epicycloid** (epi – outside). A point on the circumference of a circle rolling in the **inside** of another circle produces a **hypocycloid** (hypo – under, below). Epicycloids and hypocycloids are drawn just like cycloids but remember that the path of the centre point will be different from the path of the points marked 4 and 10. The cycloid *must* be marked from the centre line every time. The visual aid is made in exactly the same manner as the other cycloid board.

The Superior Trochoid (Superior – more than)

These curves (see Fig. 13.25) are produced when the circle rolls along a line and the path of a point fixed **outside** the circle is drawn. Imagine the path of a point on the flange of a railway train wheel. It moves below the top of the railway line and loops back.

The Inferior Trochoid (Inferior – less than)

This is the path of a fixed point inside a circle when the circumference runs on a straight line. (The path of a bicycle tyre valve.) The inferior trichoid is also illustrated in Fig. 13.25.

Gear Teeth

The cycloid is a natural rolling path and is therefore important in the design of gears. Gear teeth are shaped out of pieces of cycloids or involutes which have to mesh together without gaps and yet without jamming. Gears are not part of this book but are important examples of loci.

Figs 13.26 and 13.27 are just there to whet your appetite and to show how engineers use loci in their machines. You will not be asked to draw gear teeth at GCSE level but you may be asked at 'A' level.

Paths of Points between Circles and Lines

In Fig. 13.28 a 12 mm radius circle is near a straight line AB. A point moves so that it is always the same distance from the circumference of the circle and the line.

Method Draw OCD perpendicular to AB. Bisect CD. This gives one point equidistant from the circle and the line. Draw EF parallel to AB and 30

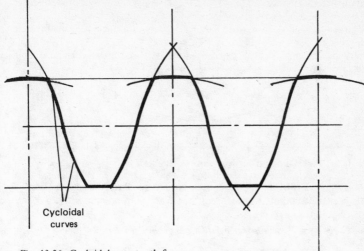

Fig. 13.26. Cycloidal gear tooth forms

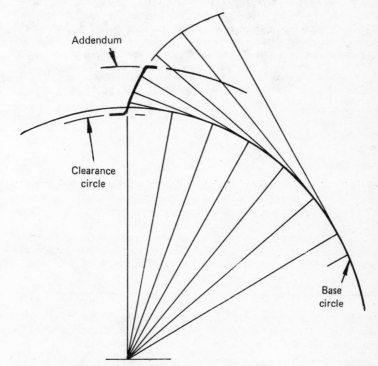

Fig. 13.27. The evolution of an involute tooth form

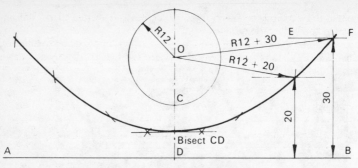

Fig. 13.28. Paths of points between circles and lines

mm from it. Draw arc centre O, radius 12 plus 30 mm to cut the line EF. This is another point. Repeat the process and join up smoothly.

The Locus of a Point always Equidistant from Two Circles

As shown in Fig. 13.29, add the same distance to both radii and strike arcs to cut each other. Repeat and join the points.

To Draw the Circle which will Touch Three Circles Internally

In this example, 'internally' means 'in the space between the circles. We

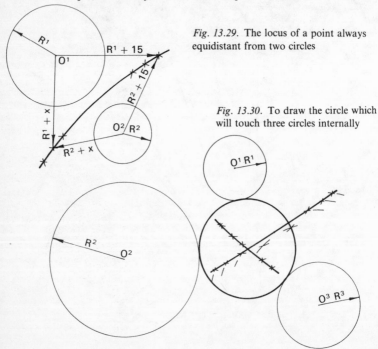

Fig. 13.29. The locus of a point always equidistant from two circles

Fig. 13.30. To draw the circle which will touch three circles internally

could draw a circle to touch the three circles and **enclose** them. This circle would touch them **externally.**

Draw the locus between two circles and then the locus between another two (see Fig. 13.30). The centre is where the loci cross. Always test this carefully before drawing the circle. It must be drawn exactly to look attractive.

Conic Sections

Important regular loci, the ellipse, the parabola and hyperbola, can be produced by cutting a cone by planes inclined in different directions. They

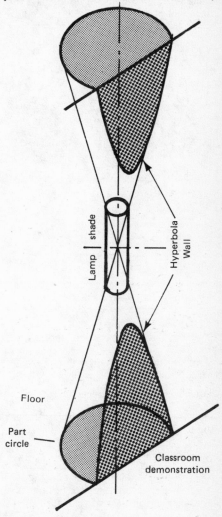

Fig. 13.31

are also called Apollonian Curves after the mathematician who first studied them. The shadows cast by the rims of a cylindrical lampshade on a vertical wall are hyperbolae (see Fig. 13.31(*a*)). By holding flat sheets (planes) at different angles across the shadow one can make ellipses, parabolas and other hyperbolae.

Fig. 13.31(*b*) shows a cone cut by a series of planes. Each one gives a different regular curve.

Fig. 13.32 is a visual aid which can be extremely useful to students.

Method of manufacture Rough turn a cone in wood. Cut it four times to give the different sections. Glasspaper these smooth and flat. Glue the cone together again with double layers of paper between each cut. Finish turn and polish. Finally, split the cone at the paper joints.

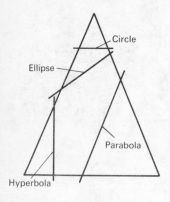

Fig. 13.32(a). Sections of a cone   *Fig. 13.32(b).* Visual aid on conic sections

Drawing Loci by Eccentricity

Eccentricity here means that the point moves so that the distance from a line and the distance from a point are always in a **fixed proportion.** The fixed point is called the **focal point or focus** and the line is **the directrix.** This sort of locus is easier than it sounds. Let us take some examples.

An Ellipse Drawn by Eccentricity

Any point on an **ellipse** is always **further** from a straight line (the directrix) than it is from a fixed point (the focus).

Fig. 13.33 shows an ellipse of $1:2$ eccentricity. This means that any point is twice as far from the directrix as it is from the focus.

Method Draw the directrix and the focal point. Draw centre line CL at right angles to the directrix through the focal point f^1. We now have to find points which are twice as far from the directrix as from f^1. Divide Cf^1

Fig. 13.33. An ellipse of 1:2 eccentricity

Fig. 13.37.

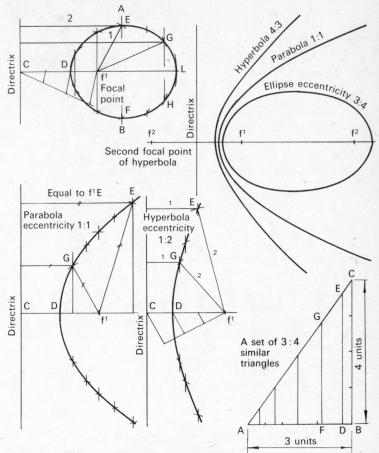

Fig. 13.34. To draw a parabola of eccentricity 1:1

Fig. 13.35. To draw a hyperbola of eccentricity 2:1

Fig. 13.36. A simple method of finding proportional sizes

into three equal parts. The second point D is on the ellipse because CD is 2 units and Df¹ is one unit. Draw AB parallel to the directrix and 30 mm away. Draw an arc of 15 mm radius and centre f¹ to cut AB at E and F. E and F are twice as far from the directrix as from f¹. Repeat with other sizes for E and F, etc. Join up.

Any point on a **parabola** is always **the same distance** from a straight line (the directrix) than it is from the fixed point (the focus).

To Draw a Parabola of Eccentricity 1:1

(See Fig. 13.34.) Repeat the construction method but with the distances from the directrix and f¹ equal to each other. CD equals Df¹ so bisect CF¹ this time.

Any point on a **hyperbola** is always **further** from the focus than from the directrix in a fixed proportion.

To Draw a Hyperbola of Eccentricity 2:1

(See Fig. 13.35.) Repeat the construction with the distance f¹E double the distance of E to the directrix. Make f¹D double DC.

A Simple Way of Finding Proportional Sizes

(See Fig. 13.36.) AB is 3 units long. BC is 4 units long. Therefore AB:BC are in the proportion 3:4. DE is parallel to BC. Therefore triangle ADE has the same angles as ABC and **must be similar to it.** This means that AD:DE are also in the proportion 3:4.

To find two lines in the proportion 3:4 using Fig. 13.36 use the following method. Take length AF. Draw parallel to BC. Then AF:FC are in the proportion 3:4.

Fig. 13.37 shows a hyperbola, a parabola and an ellipse, all with the same directrix and the same f¹.

The **ellipse** has a second focal point inside the ellipse f². This is shown but is not used in this particular construction.

The second point of the **parabola** is at infinity.

The second focal point of the **hyperbola** (if you can imagine it) goes on, in a curved infinity, comes back and hits you in the back of the neck (f² shown on the left of the directrix). If you cannot imagine it, do not bother. Just remember that it is there.

To Draw a Tangent to an Ellipse

Following Fig. 13.38 draw from the point of tangency to both focal points and bisect the angle between them. Construct the tangent at right angles.

To Draw a Circle to Touch the Ellipse

The centre is on the bisecting line in Fig. 13.38.

To Draw a Circle to Touch a Cycloid

(See Fig. 13.39.) With centre P and radius r, make an arc to cut the centre line at O. Project vertically to Q and draw QP. Extend this to form the normal, which is the centre line of the circle.

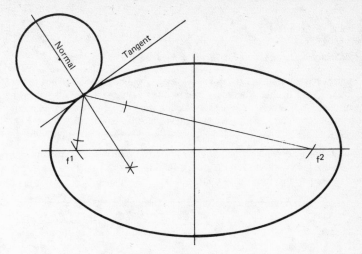

Fig. 13.38. To draw a tangent to an ellipse

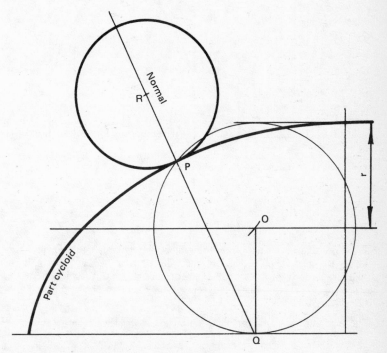

Fig. 13.39. To draw a circle to touch a cycloid

Parabolas in Architecture and Design

To Construct a Parabola to Fit Inside a Rectangle

An engineer designing a bridge knows where the sound rocks lie in the river bed. These decide where his pillars must be placed. He also knows the height of his roadway. These facts dictate the sizes of his arches. Having found his 'rectangles' he must draw the parabolas inside them (see Fig. 13.40).

Method Draw the centre line. Divide the height into four equal parts. Divide **half** the width into the same number of equal parts. Join these points as shown and join up the parabolas.

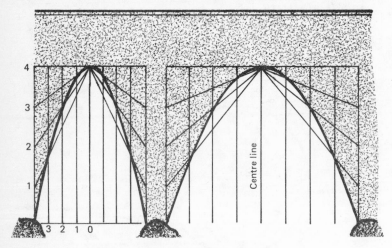

Fig. 13.40. To construct an ellipse to fit inside a rectangle

The Proportions of a Parabola

A parabola in a rectangle shows the interesting proportions shown in Fig. 13.41. If one half of the width is divided into ten equal parts and the height is divided into 100, the heights are the squares of the widths. Wren noted this in his parabola sketch for St Paul's. (See the sketching chapter – Fig. 17.18.)

The algebraic formula for a parabola is $x = y^2$. (See Computer section Fig. 19.14.)

Fig. 13.42 is a drawing, from a photograph, of the parabolic beams in the attic of the Battlo House, Barcelona (1904–6) built by Gaudi. The whole roof is sinuous and covered in coloured tiles. The colours change from left

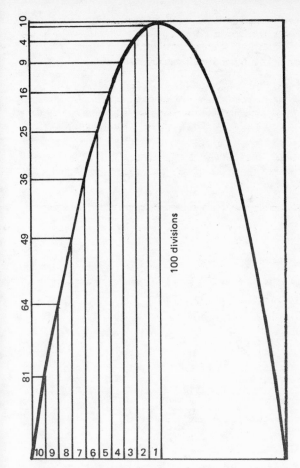

Fig. 13.41. The proportions of a parabola

to right, from dark blue to yellow, so that it looks like the back of a fish. The roof construction is an extraordinary use of parabolas, very light and strong.

A Parabolic Headlamp

Headlamps are made as parabolic bowls with the lamp in the focal position (see Fig. 13.43). Rays of light come from all sides of the lamp and strike the inside of the bowl. They are all reflected forward in a parallel beam. A headlamp always has a small black centre. This is because the light from the back of the lamp is reflected forward but hits the back of the lamp and is stopped. It cannot get past and light up its part of the beam.

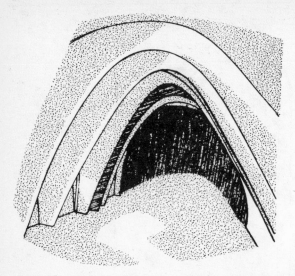

Fig. 13.42. The attic of the Battlo House, Barcelona (1904–06) by Gaudi

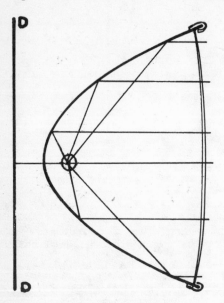

Fig. 13.43. A parabolic headlamp

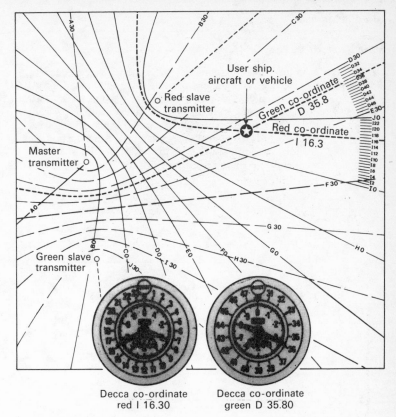

Fig. 13.44. The Decca Navigator system of location for aircraft and shipping

The Decca Navigator

This navigation system uses intersecting hyperbolic curves for radio positioning of ships/aircraft/vehicles. This position is shown as a five point star in Fig. 13.44.

A Master transmitter has two slave transmitters, Red and Green. They produce sets of radio waves which travel outwards as hyperbolae in all directions. It is perhaps easiest to think of them as curved intersecting balloons or two sets of Sydney Opera House roofs cutting through each other. They are three-dimensional, not flat. The ship or aircraft has a receiver which can read the hyperbolae and decide its own position.

In Fig. 13.42, the Red hyperbolae are drawn as solid lines and the Green ones are chain lines. The ship is at Green D 35.8 and Red I 16.3. This position would be shown on a chart.

Loci in Machines

The Action of the Shaping Machine

A shaping machine planes shavings of metal off a work piece, returns, moves slightly sideways, and planes off another shaving. (See Fig. 13.45.) It cuts going forward and this is hard work, so the machine travels forward only slowly. However, time can be saved by a quick return.

The linkwork PQR changes the rotary movement of the driving wheel to a forward and backwards (**reciprocal**) movement. The driving wheel spends 240° on the forward cut and only 120° on the return. As the wheel turns at a regular speed, the tool must return at twice the speed of the cutting,

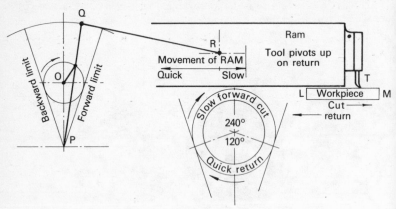

Fig. 13.45. The action of a shaping machine

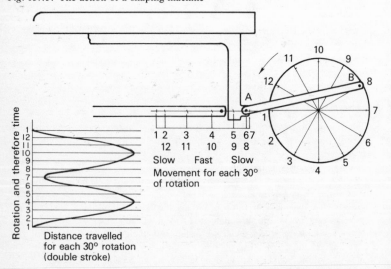

Fig. 13.46. The locus of a hacksawing machine showing changes of speed

forward cut. Two thirds of the time is spent in cutting and only one third in returning. This gives one hour forty minutes more cutting time in an eight hour shift: an intelligent use of loci.

The Locus of a Hacksawing Machine to save Jarring and Blade Damage

The driving wheel forces the saw forwards and backwards. (See Fig. 13.46.) The rotary movement of the wheel is converted into reciprocal movement. The wheel moves at a regular speed but the 30° of movement from say 7 to 8 moves the blade only a short distance. The 30° from 9 to 10 moves it much faster. The wheel turns at a regular speed so the blade must travel fast at the middle of the cut and slow up at each end. This means that the change of direction is gentle and does not shake the machine to pieces. This lengthens the life of the machine and the blades.

Machines are full of subtle details like this.

Fig. 14.1. A clockmaker's workshop. Detail from an engraving by Theodoor Galle (1571–1663).

14
Machine Drawing

A Short History of Engineering Drawing

Until the beginning of the nineteenth century, technical drawings were diagrams rather than production drawings. The way a drawbridge worked, the principle behind a barometer, the design of lock gates, could be explained by oblique sketches. A principle was illustrated. A picture gave enough information for a skilled man to make a similar article. The notebooks of Leonardo da Vinci are full of such drawings. The manufacture of accurate machine parts is another matter. Today, we buy spare parts to replace worn or broken ones. This is a very recent development.

If you lost a nut two hundred years ago, you had to go back to the man who made it for a new one, and it would have to be specially made. Even in the watch industry, where specialists mass-produced the watch parts, the making of each watch was a separate fitting job and each watch was an individual article. You could not dismantle two or three watches and interchange the parts. The engraving by Theodoor Galle (1571–1663) shows the type of hand production of the period (see Fig. 14.1).

The state of manufacturing technology was one reason why accurate drawings were not made. The second was secrecy. There was no Patent Law. Inventors hoped to keep their ideas and methods secret and so protect their livelihoods.

In the eighteenth century, the Admiralty announced a prize of £20,000 (worth untold wealth today) for the invention of a marine chronometer. With this, ships would be able to calculate their longitudes east and west of Greenwich. The calculation demanded a reliable, accurate clock. It had to be accurate to three seconds a day in seaboard conditions.

In 1759, John Harrison surpassed this accuracy but did not get his money until 1765 because the Admiralty insisted on full drawings. Harrison was very reluctant. After long nagging, he made a fine set of drawings, now held by The Worshipful Company of Clockmakers. Fig. 14.2 shows three parts of a chronometer drawn by Harrison. They are the earliest orthographic projections, where one view is **projected** from another, that I have found.

The projection lines in John Harrison's drawings are dotted. We should regard that as wrong today. Notice that he projected only two views. We have to wait for Frèzier, Monge and Marc Brunel, before we see three views projected. Harrison was drawing very early in the history of en-

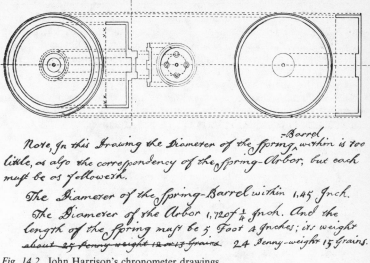

Fig. 14.2. John Harrison's chronometer drawings

Note, in this drawing the diameter of the spring-barrel within is too
little, as also the correspondency of the spring-arbor, but each
must be as followeth.
    The diameter of the spring-barrel within 1,45 inch
    The diameter of the arbor 1,72 of ¼ inch. And the
length of the spring must be 5 foot 4 inches: its weight
24 penny-weight 15 grains.

gineering drawing. He was so early that he did not know how to put on
measurements, and had to write them out. He was in good company. Noah
did not understand drawings either. God had to give him a specification,
not a drawing, for the Ark. 'The length of the ark shall be three hundred
cubits, the breadth of it fifty cubits and the height of it thirty cubits.'
There was no drawing.

Gaspard Monge, who invented what he called 'Descriptive Geometry',
was born near Dijon, in France, in 1746. He went into the French Army as
a draughtsman but was not rich enough to become an officer. When asked
to calculate the volume of earth to be moved to build a fort, he drew it as
a set of three-dimensional blocks and calculated the volumes. He worked it
out so fast that his teacher did not believe him. These calculations had
always before been worked out laboriously by arithmetic, in wheel-
barrowsful. When they realised the advantages of the method, the French
Army, always at war with the neighbours, kept orthographic projection as
a state secret for thirty years.

After the French Revolution, Monge taught the method publicly. It
allowed a three-dimensional object to be drawn accurately on a two-
dimensional piece of paper, with each view projected from the other views.

Marc Isambard Brunel was another Frenchman, but unlike Monge on the Royalist side. He was a naval officer and escaped from France, first to America and later to England. He designed the famous block-making machinery at Portsmouth Dockyard. His drawings, published in Ree's *Encyclopaedia* in 1819, were the first three view orthographics published in England. Frèzier had published some earlier in France.

Gradually, the knowledge of orthographic projection spread and, when mass-production changed the world, the method of drawing for it had been worked out. Today we are familiar with technical drawings and have agreed how to draw them. There is, however, a danger.

Before the end of the eighteenth century, there were few engineers and they made sketches for their own work. With the beginning of mass-production, draughtsmen tended to become a separate group, rather 'superior' to the artisans who built the machines. This separated some of them from the practical work and the 'feel' of machines. Not knowing the practical problems, today, a draughtsman can draw 'rubbish' which does not work or cannot be made.

A high-speed letter sorting machine was designed to be worked by an electric motor. The draughtsman received the design engineer's sketches. Instead of one motor, he substituted four small ones, each working a different part. The drawings were pretty. Nobody noticed the change. When the machine was switched on, the motors turned at slightly different speeds and, instead of sorting the letters, the machine tore them up like confetti.

Draughtsmen, especially in a world where materials and methods change so rapidly, need to be alert, imaginative but cautious people. The evil that men do can live after them.

Sectioning

A section can often make a drawing clear without taking extra space or requiring more views. **The following Sectioning Conventions are important** (see Fig. 14.3)

1 **A Simple Section** The box is cut by section line AA. The arrows point in the direction you are looking. The section has been drawn on the right. This is because the projection is in First Angle. If it was in Third Angle, the section would be drawn on the left.
2 **A Part Section** A piece has been broken away to reveal the section of an important part of the box. This saves space.
3 **Sectioning of Webs** Webs are the stiffening plates in castings. We never section webs, as this would make the metal look too solid and give the wrong impression. Part 3a is right and part 3b is wrong.
4 **Revolved Sections** The shape of the spoke has been turned through 90° and cross hatched. The sections of the bars have been turned in the same way.
5 **A Rotated Section** The section line goes down the centres of the two spokes. The top section can be seen correctly. It would be difficult to draw the lower section properly as it is sloping away and **it would be of no value to an engineer**. Instead, the section has been drawn as if it had been turned vertically.

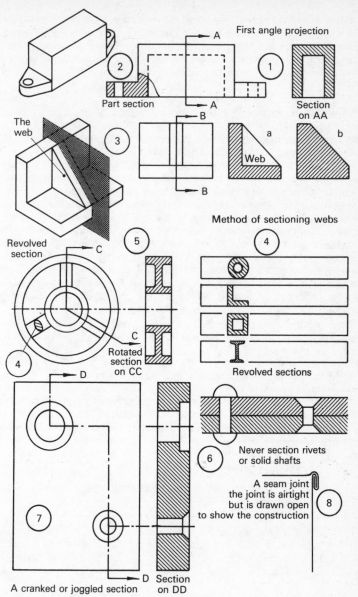

Fig. 14.3. Sectioning conventions

6 **Sectioning of Solids and Tubes** Never section solid bars, rivets, shafts, etc. because this would not tell us anything extra. We section hollow shafts; bars, castings etc. of special sections; and other objects **where sectioning will make the drawing clearer**.

7 **Cranked or Joggled Sections** When the important parts of a drawing are not in line, the section line can be cranked to go through them as shown.

8 **Sectioning of Sheet Metal** The lines are drawn with gaps between them to reveal the construction. Everyone knows that, in fact, the sheets are hammered together and the joint may even be watertight.

Introduction to Machine Drawing

In chapter 3 we began the study of orthographic projection and drew some small engineering parts in First Angle and in Third Angle projection. You should look back at this section to revise the differences between the two. Remember that elevations are **always** side by side. In First Angle projection the plan is **below** and in Third Angle, the plan is **above**.

Machine drawings must always be very clear and as simple as possible. Different parts of an article may be manufactured by different people, perhaps in different factories and even in different countries. Those making the parts will not be able to talk to each other and discuss problems, so the drawings must include every necessary fact, clearly stated. Try to put yourself in the position of someone who has never seen the article to be made and may have no idea how it works or how his part fits into the scheme of things. If your drawings are not clear, you will be inviting trouble.

Layouts

Before starting any drawing, plan the layout. Fig. 14.5 shows a layout for Fig. 14.4. Measure the maximum length, width and height. Set out blocks in the correct positions making sure that your elevation is on the correct side. Leave plenty of room for measurements. Then calculate how the blocks will fit neatly.

Every machine drawing must have a **border** of at least 10 mm **all round it**. The border is the proof that you have a complete drawing. **Never use a drawing without a border all round**. The drawing may have a vital section missing and you may cause a worthless item to be made.

Small Engineering Drawings

Fig. 14.4 is a **simple bearing bracket** for carrying a shaft. A replaceable phosphor-bronze bearing (not shown) fits inside the hole to take the wear of the shaft. The section line AA has been joggled to show the detail of both types of hole.

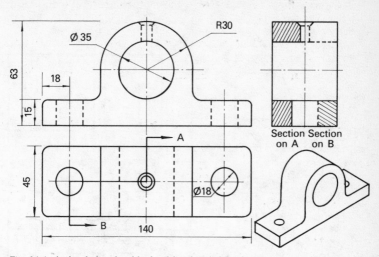

Fig. 14.4. A simple bearing block with a joggled section

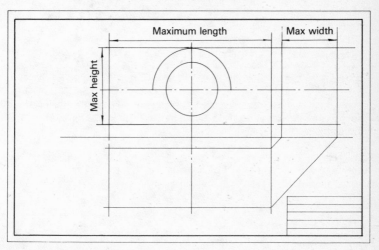

Fig. 14.5. A layout for Fig. 14.4

250 *Graphic Communication*

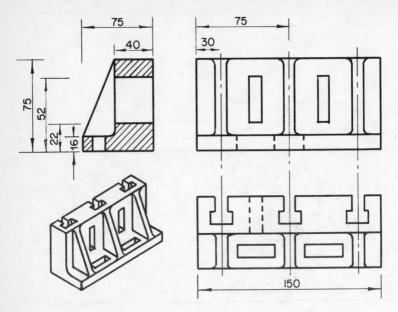

Fig. 14.6. A planing machine bracket

A Planing Machine Bracket

Plan a layout for a front elevation, a plan and a sectional end elevation on the right of the elevation. Add a title block, scale and six important measurements. (See Fig. 14.6.)

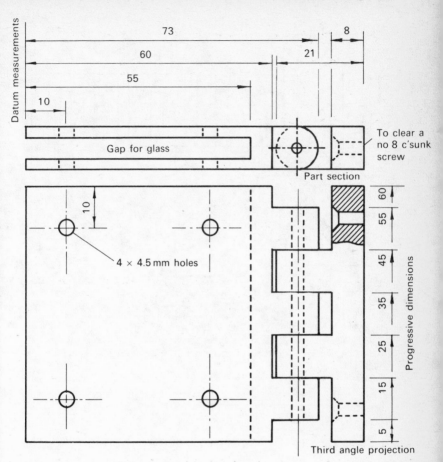

Fig. 14.7. A shower screen hinge

A Shower Screen Hinge

Copy Fig. 14.7 but leave more room for the measurements. The drawing shows **datum measurement** and **progressive measurement**. Compare the two and notice that progressive measurement takes less room but is not quite so clear. The **part section** shows important detail without taking extra room.

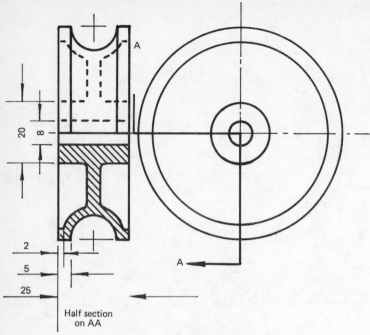

20

8

2

5

25

A

A

Half section
on AA

Fig. 14.8. A plastic pulley wheel – full size, Third Angle projection

Fig. 14.9. A door turn with a sunken handle

A Plastic Pulley Wheel

An example of a drawing with one half in section is given in Fig. 14.8. We get two drawings in the space of one.

A Door Turn with a Sunken Handle

In Fig. 14.9 the sectional end elevation shows four different parts. Each one is sectioned differently.

Fig. 14.10. An exploded isometric drawing of a sunken handled door turn

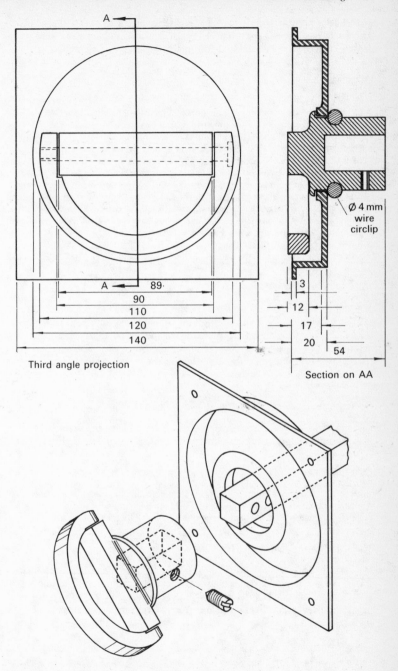

A

A ← 89·
90
110
120
140

Third angle projection

∅ 4 mm
wire
circlip

3
12
17
20
54

Section on AA

The door handle is shown in an **exploded isometric** drawing in Fig. 14.10. The parts have been separated in the directions in which they were withdrawn. The drawing explains, without words, how to re-assemble it.

Questions

1 Draw Fig. 14.4 full size.
2 Draw an exploded isometric projection of Fig. 14.7. Do not add measurements. They are difficult to fit on to isometric projections.
3 Draw Fig. 14.8 full size but with both sides sectioned.
4 There are rather a lot of circles for Fig. 14.8 to be drawn in isometric projection. Instead, draw it in oblique projection with the circles drawn with a compass.
5 Draw Fig. 14.6 with the sectional end elevation on the left. Add a full end elevation on the right. Also add a scale, the projection, six important measurements and a title. (Examination questions often ask for scale, etc. to be entered on the drawing. This means that marks have been set aside for these details. If you do not put them in, you cannot have the marks. **Even if you have to leave a drawing unfinished, put in the details you are asked for, and earn these marks.**)
6 Draw the body only of the door turn (Fig. 14.10) in isometric projection. Think of it as a set of very short cylinders.

A Lever Door Handle

A great deal of information is packed into Fig. 14.11. Notice the conventional drawing of a tension spring and the way the handle is held on the door plate by being swaged **over** the washer. The door handle can be assembled (with a different lever) to open the other way. The spring is then held on the other peg and the other lug on the stop washer, presses against the top stop.

An Exploded Isometric Drawing of the Handle Assembly

In Fig. 14.12 notice that every part has been pulled out along the centre line. The drawing was made by using an ellipse template (see Fig. 1.8).

Questions (continued)

7 Draw the three views in Fig. 14.11. Add a title block, the scale, six important measurements and the First Angle or Third Angle symbol.
8 Draw any one part of the handle assembly in isometric projection.

A Plastic Spring Clip with Hook

Fig. 14.13. deals with a tangency question. You may need to revise Figs 2.76–2.81 in Chapter 2. Start by drawing the centre lines and setting out all circles from them. Then find the tangency centre 0^1 as shown. Line in neatly and **leave in all construction**. Add six important measurements, the projection and title.

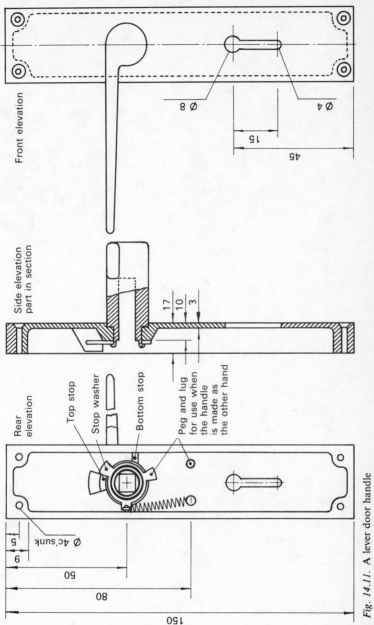

Front elevation

8 Ø

4 Ø

15

45

Side elevation
part in section

17
10
3

Rear
elevation

Top stop

Stop washer

Bottom stop

Peg and lug
for use when
the handle
is made as
the other hand

Ø 4 c'sunk

5

9

50

80

150

Fig. 14.11. A lever door handle

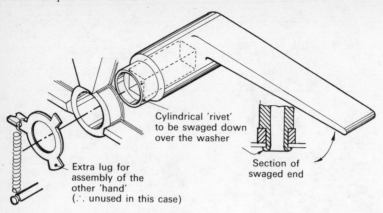

Fig. 14.12. An exploded isometric projection of the lever handle assembly

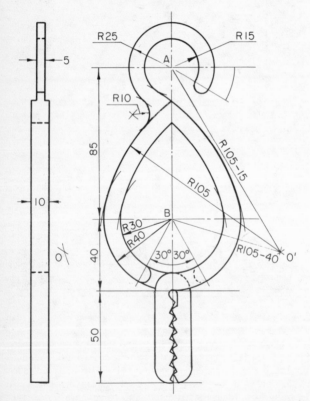

Fig. 14.13. A plastic spring clip with hook – First Angle projection

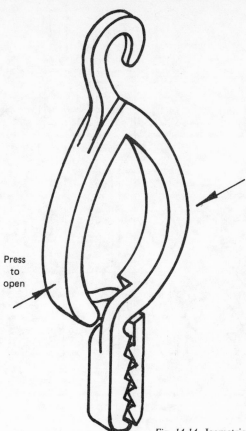

Press
to
open

Fig. 14.14. Isometric projection of the spring clip

Isometric Projection of the Spring Clip

Fig. 14.14 is a very difficult drawing. Try it if you like but it is at 'A' level standard.

Making Measured Drawings

It is important to learn by drawing real things and working out the best way of showing them clearly. A drawing is only a way of giving the 'reader' a set of information. If the drawing does not give the information clearly and simply, it is a nonsense. Copying other people's drawings can be a good way of learning how to use instruments and to set things out, but please do not copy slavishly. Remember the fable of the tailor who was asked to copy a coat. He copied everything perfectly, including the tear in the sleeve.

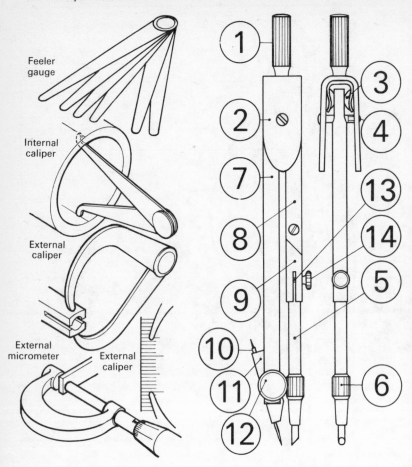

Feeler
gauge

Internal
caliper

External
caliper

External
micrometer

External
caliper

Fig. 14.15. Measuring instruments   *Fig. 14.16.* A geared compass

Measuring Tools

Fig. 14.15 shows four typical measuring tools used in engineering.

Feeler Gauges Used to measure fine gaps such as tappet openings. Each leaf is a different thickness and combinations of leaves can be used to measure any particular gap.

   Internal and External Caliper Opened to fit objects and then the tips are measured against rulers.

   External Micrometer Used for measuring sizes very accurately. One could add a radius gauge, a vernier caliper, a height gauge, protractor, etc. Any of these may be necessary if you need to draw an object accurately but, in many cases, simple rules and calipers are enough.

Fig. 14.17. A detail sketched on squared paper

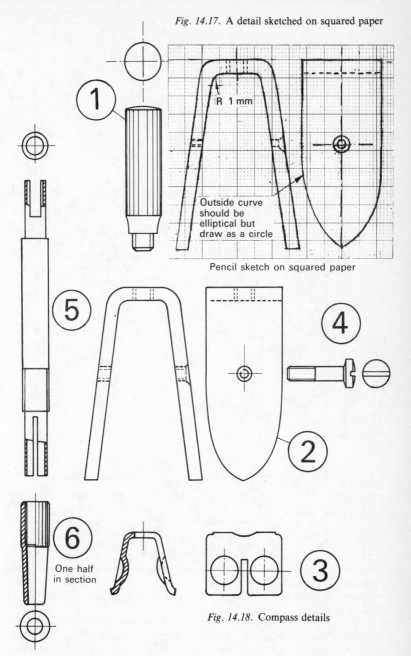

R 1 mm

Outside curve
should be
elliptical but
draw as a circle

Pencil sketch on squared paper

One half
in section

Fig. 14.18. Compass details

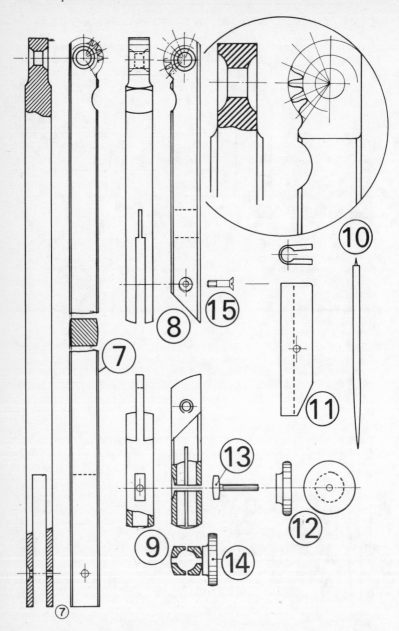

Fig. 14.19. Compass details

The Assembled Compass

Suppose you decide to draw a compass to scale, with measurements. This would be a large task but you need not tackle it all at once. A few parts would make a start.

Fig. 14.16 shows an **assembled compass** with each part numbered. Although this drawing appears first in the book, it was the last to be drawn. It is much easier to draw the detail drawings first and then put them together.

Method Make freehand drawings on squared paper of each part as in Fig. 14.17. The squares may give the measurements accurately enough but exact sizes are generally added.

From the squared paper drawings, make detail drawings of each part in turn (see Figs 14.18 and 14.19). Number and name each part and add measurements. They have been left out of these examples because of space, but you should spread out the drawings and add all necessary measurements. Save space by adding half-sections, part-sections, etc. where they are helpful.

Each part can make an interesting little drawing requiring two or three views. Some parts, such as the geared head shown in part number 8, may need to be enlarged.

If you get far enough, make the assembly drawing from your detail drawings, in the same way as Fig. 14.16 was made.

A compass is a very elaborate example. Start with much simpler objects from around your house and workshop. Woodwork sash cramps are ideal for this type of drawing **but they must be measured from real ones.**

Questions

Take a small object of a simple kind, like Figs 14.4 or 14.7. Make a set of drawings as follows:
1 A freehand sketch on squared paper.
2 A finished orthographic projection.
3 An isometric drawing
4 An axonometric drawing.
5 Trim and mount all of these on a piece of coloured paper. These sets make attractive wall displays and parts of the course work entries for some exams. Some examinations give as much as 30 per cent of the marks for work done during the course. These marks can make the difference between a good pass and scraping through. Check up with your teacher on the rules for your particular examination. This sort of drawing set can be repeated as often as you like.
6 Another attractive exercise is to make a set of drawings of the table or drawing stand at which you are working.

A Bicycle Pedal

These bicycle pedal drawings are made from a set of drawings kindly supplied by Sturmey-Archer Ltd but I must point out that these are my much simplified versions of the originals.

Fig. 14.20 shows the assembled pedal in plan and elevation with a sec-

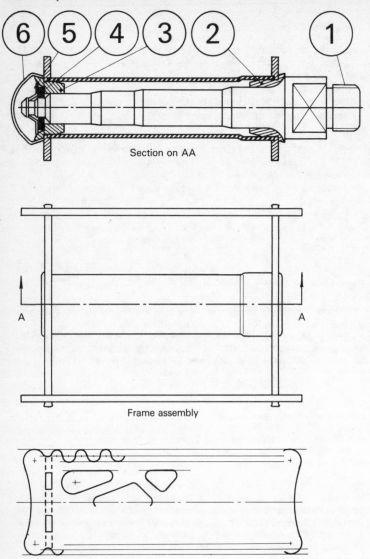

Section on AA

Frame assembly

Fig. 14.20. A bicycle pedal
By permission Sturmey-Archer Ltd

tional view on AA. Notice that the shaft is not sectioned, as we never section solid shafts. The parts are numbered so this is a key view for the later detail drawings (See also Fig. 14.21.)

Some Parts of the Pedal			
Item	Decription	No. off	Material
1	Spindle RH/LH	1	M/S
2	Bush	1	Bronze
3	Bush	1	Bronze
4	Washer	1	M/S
5	Ratchet plate	1	Spring steel
6	End cap	1	M/S
7	Pedal side plate	1	M/S

Fig. 14.21.

Detail Drawings

Figs 14.21 and 14.22 have measurements giving top and bottom limits. Any machine part which is smaller than the small measurement or larger than the big one, will be rejected as scrap. The parts will be tested with Go and No-Go Gauges. These are specially made measuring gauges. The part must go in at one end and not go in at the other.

Engineers sometimes give a middle figure with a plus and minus figure instead of the top and bottom ones.

For example: $\frac{12.77}{12.65}$ could be written 12.71 ± 6

Shafts fitting into holes need a different type of tolerance. If the shaft was oversize and the hole undersize, the shaft might not enter. Therefore the shaft has an upper limit and the hole has a lower limit (see Glossary).

Questions

Do not attempt all these at once. Make some detail drawings from Figs 14.18–14.20. Put each drawing in a block and then draw the next one. When you make your drawings, give yourself plenty of room. Spread out the measurements and draw at least four times the size shown in the book.

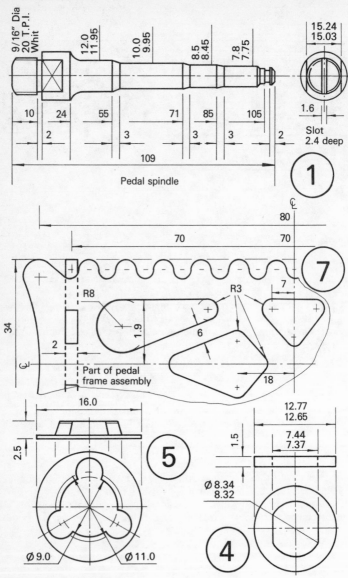

Fig. 14.22. Pedal details

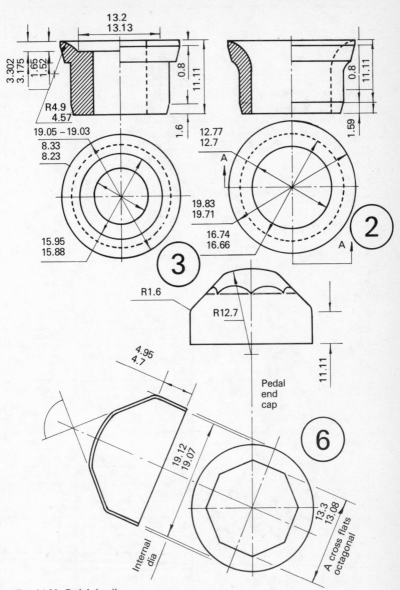

Fig. 14.23. Pedal details

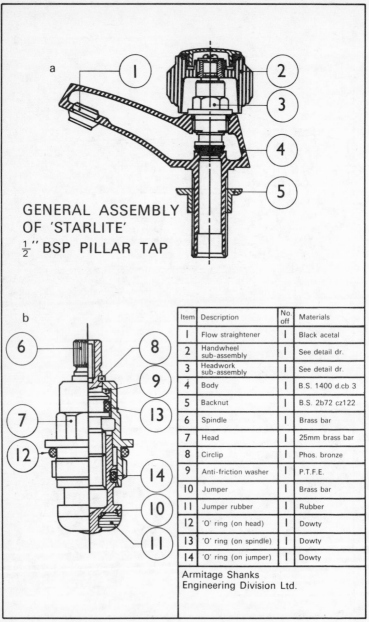

Item	Description	No. off	Materials
1	Flow straightener	1	Black acetal
2	Handwheel sub-assembly	1	See detail dr.
3	Headwork sub-assembly	1	See detail dr.
4	Body	1	B.S. 1400 d.cb 3
5	Backnut	1	B.S. 2b72 cz122
6	Spindle	1	Brass bar
7	Head	1	25mm brass bar
8	Circlip	1	Phos. bronze
9	Anti-friction washer	1	P.T.F.E.
10	Jumper	1	Brass bar
11	Jumper rubber	1	Rubber
12	'O' ring (on head)	1	Dowty
13	'O' ring (on spindle)	1	Dowty
14	'O' ring (on jumper)	1	Dowty

Armitage Shanks
Engineering Division Ltd.

Fig. 14.24. The 'Starlite' ½″ BSP pillar tap with numbers and title block
    (*a*) General assembly
    (*b*) Sub-assembly of ½″ performance headwork
By permission of Armitage Shanks

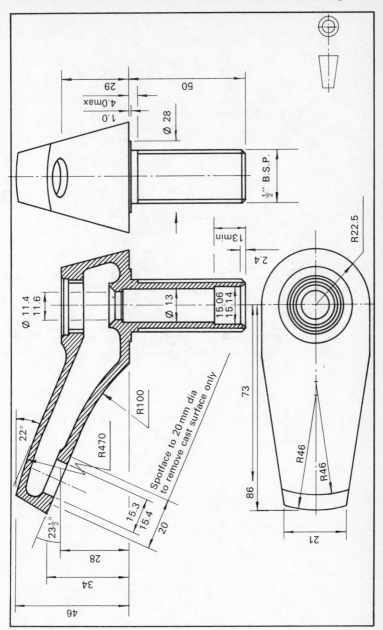

Fig. 14.25. Machining details of $\frac{1}{2}''$ tap body

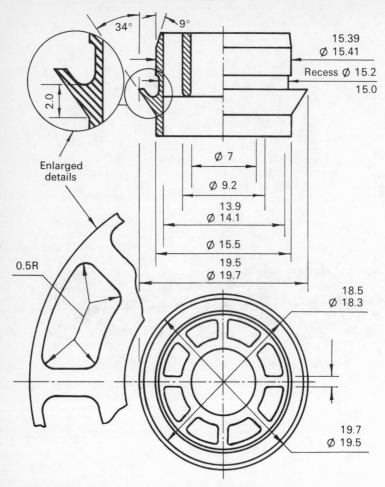

Fig. 14.26. Detail drawing of flow straightener for ½″ tap

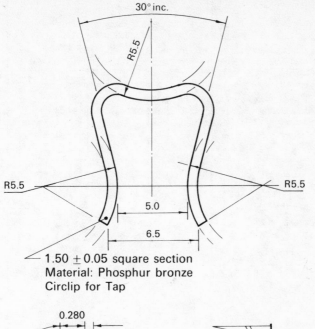

1.50 ± 0.05 square section
Material: Phosphur bronze
Circlip for Tap

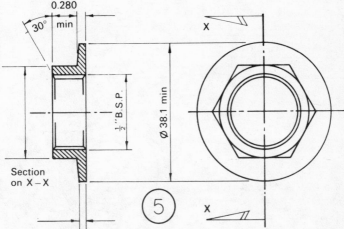

Fig. 14.27. Detail drawing of back nut

The 'Starlite' ½″ BSP Pillar Tap

The 'Starlite' ½″ BSP pillar tap drawings (Figs. 14.25–14.29) were made from drawings kindly supplied by Armitage Shanks. Draw these as you did the pedal drawings.

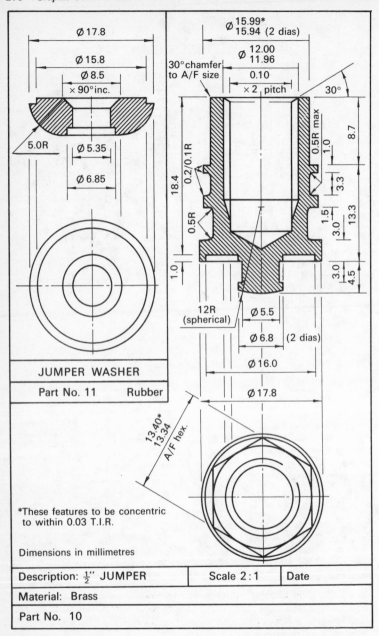

Ø 17.8

Ø 15.8

Ø 8.5
× 90° inc.

5.0R

Ø 5.35

Ø 6.85

JUMPER WASHER

Part No. 11   Rubber

Ø 15.99*
15.94 (2 dias)

Ø 12.00
11.96

30° chamfer
to A/F size

0.10
× 2 pitch

30°

0.5R max

8.7

1.0

18.4

0.2/0.1R

3.3

0.5R

1.5

13.3

3.0

1.0

3.0

4.5

12R
(spherical)

Ø 5.5

Ø 6.8   (2 dias)

Ø 16.0

Ø 17.8

13.40*
13.34
A/F hex.

*These features to be concentric
to within 0.03 T.I.R.

Dimensions in millimetres

Description: ½″ JUMPER | Scale 2:1 | Date

Material: Brass

Part No. 10

Fig. 14.28. Circlip for tap

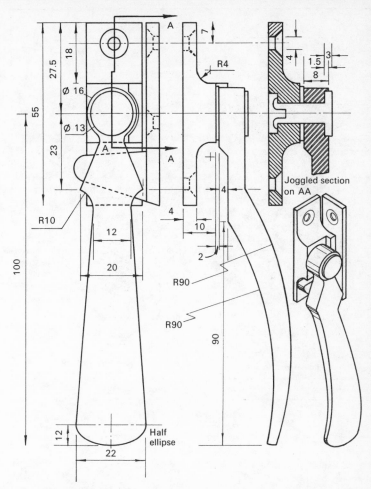

Fig. 14.29. Detail drawings

A Casement Window Closer

Fig. 14.29 has some interesting tangency problems which can be drawn simply.

Front Elevation Draw a half ellipse at the bottom and the circle radius 10 mm. Join with a rule. It is not necessary to construct elaborate points of contact in a machine drawing like this. Notice also that section AA is joggled to go through the two centres.

    End Elevation Find the four end points of the curves of the handle. Strike arcs of 90 mm radius to find the centres of the two quarter circles.

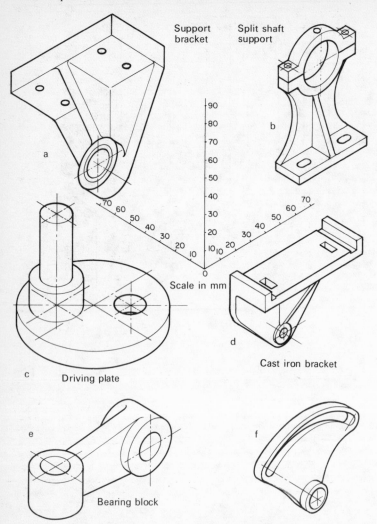

Support bracket

Split shaft support

b

90
80
70
60
50
40
30
20
10

70
60
50
40
30
20
10

a

70
60
50
40
30
20
10 10

0

Scale in mm

c

Driving plate

d

Cast iron bracket

e

f

Bearing block

Fig. 14.30. A casement window closer – First Angle projection

Isometric Projection The handle is difficult and well above GCSE level examination standard. Instead, draw the two screwed blocks in isometric projection four times the size shown in the book. They are at GCSE level standard. Then, if you like, try the handle.

Small Engineering Parts

The isometric drawings shown in Fig. 14.31 are typical of those found in

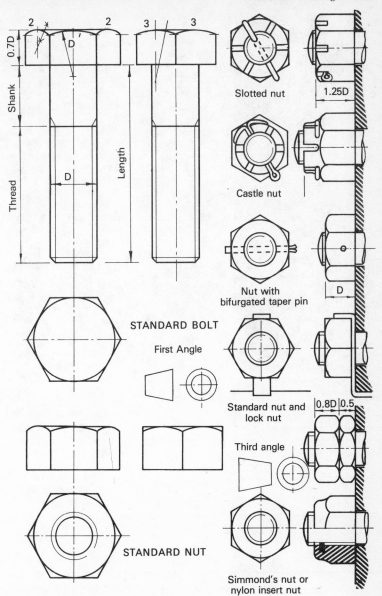

Fig. 14.31. Small engineering parts

Measurements must be taken in the three scale directions only. All other dimensions are distorted in isometric projection.

Fig. 14.32. The standard ISO nut and bolt

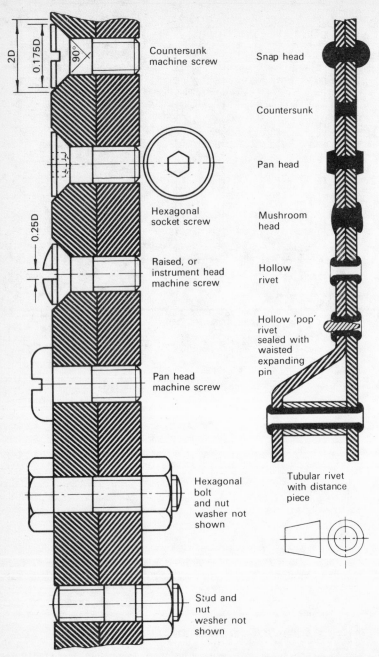

Countersunk
machine screw

Snap head

Countersunk

Hexagonal
socket screw

Pan head

Raised, or
instrument head
machine screw

Mushroom
head

Hollow
rivet

Pan head
machine screw

Hollow 'pop'
rivet
sealed with
waisted
expanding
pin

Hexagonal
bolt
and nut
washer not
shown

Tubular rivet
with distance
piece

Stud and
nut
washer not
shown

Fig. 14.33. Locking nuts   *Fig. 14.34*. Standard screws bolts and rivets

engineering drawing questions. The examiner could use them in a dozen different ways. These are examples:

1 Draw a front elevation, a plan and a side elevation looking from the right. State which projection you have used.
2 (a) Draw a front elevation and plan in Third Angle projection.
   (b) Add a sectional elevation to the left of the first elevation.
3 (a) Draw an elevation in First Angle projection.
   (b) Draw a plan.
   (c) Draw an end elevation looking from the right of (a).
   (d) Draw a sectional end elevation looking from the left of (a).
4 Draw a sectional elevation in Third Angle projection.
   Project a plan.
   Project a sectional end elevation looking from the left of the sectional elevation.

The examiner could also ask for isometric drawings looking from a different view point, or for oblique projections.

You should draw a number of the questions posed, noticing the **absolute importance of the wording of the question**. If you make an excellent drawing of the wrong question, the examiner **cannot** give you any marks. He may sympathise but he **cannot** give you marks.

The Standard ISO Nut and Bolt

Nuts and bolts are made by specialist firms. No normal engineer would think of making a standard nut or bolt: it could be bought much cheaper off the shelf.

Bolts and nuts are made to International Standards Organisation sizes (ISO) and are drawn in proportions of the shank diameter D (see Fig. 14.32). There is one standard shape which is enlarged or reduced.

To Draw a Bolt or Nut given the Distance across the Flats (A/F)

Draw a circle equal to the A/F size. Construct a hexagon round the circle. Project the front and side elevations. Mark the thickness. Nuts and bolts have the corners chamfered at 30° for safety. Draw the centre curve first with R = D. This gives the level of all the other curves. Bisect the other faces and find centres by bisecting the arcs as at 2.

Locking Nuts

Fig. 14.33 shows a series of locking devices for preventing nuts being undone by vibration. This is an everlasting danger in aircraft, motor cars, etc. These are some of the many varieties in everyday use. Make a collection of locking nuts yourself.

Engineers' Screws, Bolts, Studs and Rivets

The differences between these must be clearly understood (see Fig. 14.34). Each has its proper function. Try to find examples of each and work out why a particular one has been chosen.

15
Surveying

Equipment

Surveyors have to measure and draw fields, roads, building sites, archaeological digs, etc. for planning and recording. This chapter deals with simple surveys of fairly level sites. Hilly ones are more complicated, requiring the use of theodolites. These are revolving telescopes which can measure angles both sideways and up and down. Simpler instruments, called dumpy levels in the trade, measure only angles of turning.

Keep your eyes open and you may see surveyors today, in the street, using the modern optical instruments which read directions, distances and up and down angles, on digital read-outs. They are very advanced and expensive.

Fig. 15.1 shows Humphrey Repton's trade card. He was the great landscape 'Improver' (1752–1818). He re-designed farms as gentlemen's estates,

Fig. 15.1. Humphrey Repton's trade card, showing a surveyor using a theodolite

establishing an avenue of trees, or building artificial mounds with brand
new, romantic ruins on the tops. His trade card shows him using a theodo-
lite to set out a park, while labourers widen the local brook into a lake
which apparently continues, past the trees, for ever. No doubt it really
stops dead and the view behind, is ugly, with a factory or coal-mine where
the client's money comes from.

You will need the following instruments for a simple survey:

Metal surveyor's chain (22 ft) and/or a measuring tape 15 m (50 ft) or 30
    m (100 ft).
Ranging rods or broomsticks for lining up straight lines.
Pegs and a mallet.
String on a winder.
A clipboard, pencils and paper in a plastic bag. The latter should be large
    enough to allow the hand and forearm free play to write inside the bag
    during rain.
A surveyor's notebook. A real surveyor's notebook has a hard back and
    the pages are numbered as they can become important documents. For
    practice work we can draw our own sheets.
Normal drawing instruments back at home.
Good feet.

Setting out a Right Angle with String

This is often important. It is a very old piece of geometry. Each year the
River Nile flooded, wiping out the field boundaries. When it subsided, the
fields had to be set out anew. This need to re-survey the land every year,
so that each farmer got his fair share, caused the Egyptians to be very
interested in simple geometry. If a field corner was not a right angle, one
farmer gained a few feet by the end of the field and his neighbour lost
some. The surveyor had two people watching every move.

Fig. 15.2 shows an Egyptian surveyor and his two men carrying the
knotted cord which they used for measuring and setting out right angles.

Fig. 15.2. An Eygptian surveyor and his men surveying a field. From the tomb
of Amenhotepe-si-se at Thebes. (From Davis, *The Tomb of Two Officials of Tuthmosis.*)

A triangle with sides in the proportions 3:4:5 must contain a right angle. This is the Egyptian Osiris triangle. Osiris was their god of fertility and rejuvenation.

Method

(a) Take a piece of string knotted at twelve equal distances (see Fig. 15.3).
(b) Peg down the two ends together.
(c) Pull the point 3 taut and peg down.
(d) Pull point 7 so that both strings are taut and peg down.
    The 0, 3, 7 corner must be a right angle (Pythagoras).

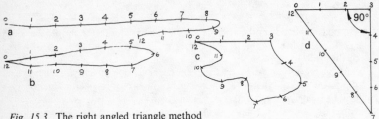

Fig. 15.3. The right angled triangle method

Sample Surveys

Fig. 15.4 shows a simple field to be surveyed. The surveying is done in two stages.

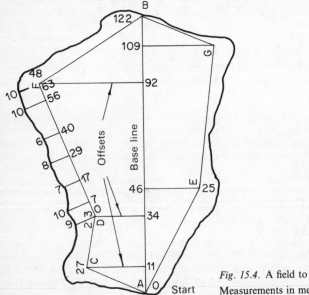

Fig. 15.4. A field to be measured

Measurements in metres

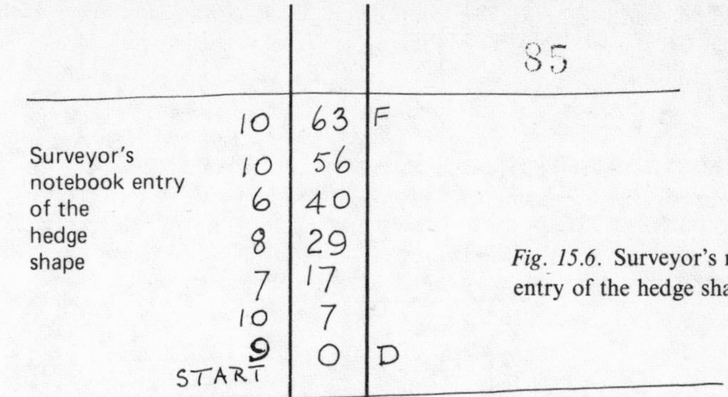

85

10	63	F
10	56	
6	40	
8	29	
7	17	
10	7	
9	0	D
START		

Surveyor's notebook entry of the hedge shape

Fig. 15.6. Surveyor's notebook entry of the hedge shape

	122	
	109	33
48	92	
	46	
23	34	
27	11	
START	0	

Surveyor's notebook entry of the field shape

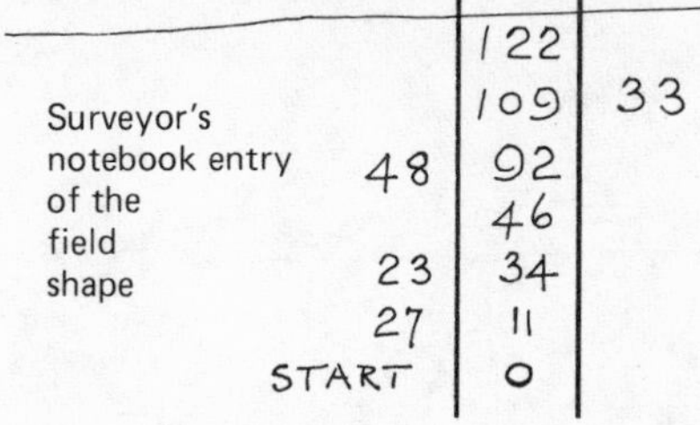

Fig. 15.5. Surveyor's notebook entry of the field shape

Stage 1 The shape OACDFBGEO is surveyed to give the **rough** shape of the field.

Stage 2 Then the **exact** shape of the hedge is set out from the edges of shape 1.

Method Drive in three poles in a **straight** line AB, sighting them carefully. Peg out a string 0–122 along the poles. This is called your **base line**. (The base line is also called the chain line or datum line.)

A Surveyor's Notebook

A surveyor's notebook has a double line printed along the centre of its length (see Figs 15.5 and 15.6). The book is held so that the double line is upright and the measurements are entered from the bottom to the top. The base line measurements go **between** the pair of lines, with 0 at the bottom and 122 above.

Offsets are lines set out to left or right of the base line, at right angles. They go to important points near the hedge.

To Set out Point C

Start at A (as shown in Fig. 15.4) and walk along the base line to a convenient point roughly at right angles to C. It is 11 metres from A. Write 11 between the parallel lines in your book (see Fig. 15.5). On the ground, set out a right angle towards C. Measure from 11 to C (27 m). C is on the left

Fig. 15.7. An estate in Richmond belonging to John Spooner and John Roper 8. 5th 1779 – the field sketchbook drawing

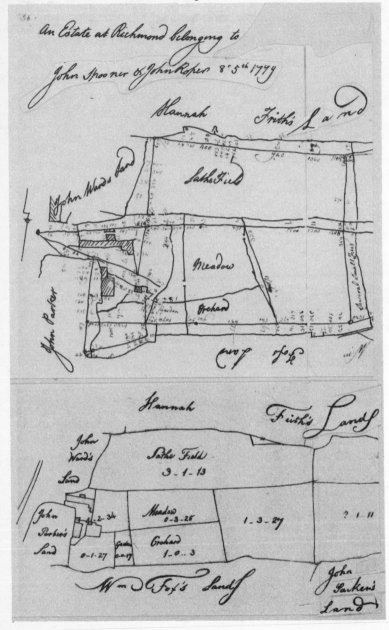

Fig. 15.8. The finished estate map

of the base line, so enter 27 to the left of the double line in the book. Set out 34–D and 92–F in the same way. Corner E is on the right so enter 46 between the lines and 26 on the right.

Each set of measurements between the parallel lines in your book is called a **traverse**.

Finding the Hedge Shape

Example Walk from D to F. The hedge is on your left. DF becomes the next base line so its measurements are written down between the lines in your notebook (see Fig. 15.6). Put 0 at the bottom between the lines. Set out offsets at convenient points (7, 17, 29 in this case). Measure the offsets and enter them on the left in the notebook. Complete the hedge around the whole field, F–B, B–G, etc. Start from 0 again each time you turn.

Back at home, plot the survey accurately from your notebook figures.

An Estate in Richmond belonging to John Spooner and John Roper 8' 5th 1779

Fig. 15.7 shows a piece of a survey notebook of the village of Richmond, on the south of Sheffield, made in 1779. Three years after the American Declaration of Independence, a surveyor was quietly walking round the fields, surveying them in order to make a map of the estate. He would have had a couple of men with a chain and pegs, to lay out the marking strings, and probably a young apprentice to carry his inkwell. You can see his chain lines (single) and the offsets, on his sketch map. There is room in this book for only one end of his map but this shows plenty of detail.

The Finished Survey

Fig. 15.8 shows the finished survey of the same land, in beautiful copper-plate writing, full of flourishes, and the areas calculated in acres, roods and poles. (1 acre = 4 roods; 1 rood = 40 poles. A pole was $5\frac{1}{2}$ yards but these are square poles, $30\frac{1}{4}$ square yards.)

Fig. 15.9 shows a field rather like that in Fig. 15.4 but with a stream running across it. We will survey this one from point A but without using a base line. Instead, we will measure radiating lines from A. Later we will draw the field by constructing a network of triangles.

Fig. 15.10 shows a surveyor's notebook entry for this piece of surveying.

Method of surveying the field Start from A as 0. Straight ahead is B at a distance of 150 m. The two ends of the bridge across the stream are at 75 and 78 (There is no point in putting m for metres each time.) Radiating from A are the points C, D, E, etc. and their distances from A. The double line shows the finish of that traverse. Above are the distances B to C, C to D, etc.

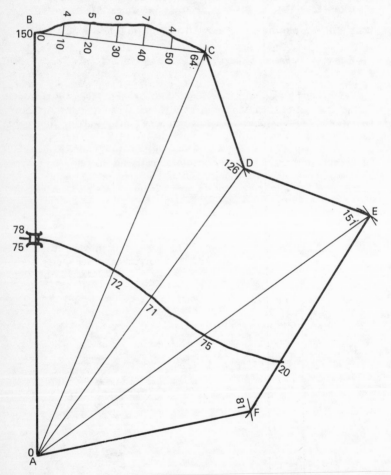

Fig. 15.9.

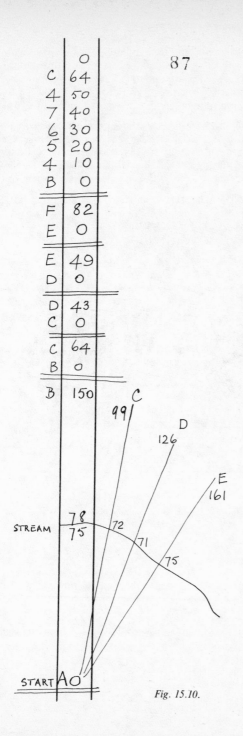

87

	0
C	64
4	50
7	40
6	30
5	20
4	10
B	0
F	82
E	0
E	49
D	0
D	43
C	0
C	64
B	0
B	150

STREAM

START A 0

Fig. 15.10.

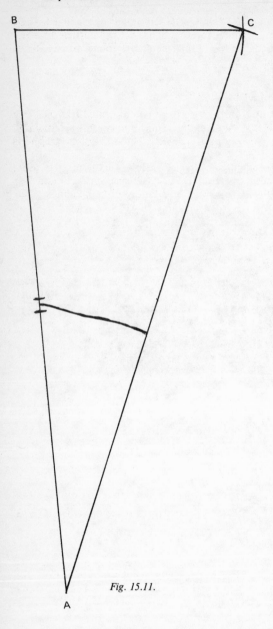

Fig. 15.11.

Setting out the survey Fig. 15.11 shows the setting out of the first triangle ABC. Strike arcs from A and B to cross at C. Set out further triangles similarly. Set out the hedge B to C using the figures found on page 87 or the surveyor's notebook.

An Archaeological Site with Four Buildings

This simple survey (see Fig. 15.12) consists of two measured triangles. Peg out the strings AD and BC. Measure the distances to the crossing point 0. Measure AB and CD. Make a measured sketch (the base line method is not used here).

The buildings are set at different angles and the survey must measure these. A triangle has been set out at D. The sides measure 4, 5 and 7 so any 4:5:7 triangle will give the correct orientation of the building.

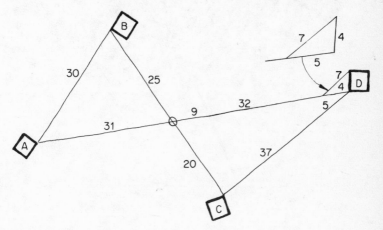

Fig. 15.12. Four buildings on an archaeological site

A Cliff Path

Some areas are difficult to survey. A cliff path is typical of such areas. A twisting cliff-top path is shown in Fig. 15.13. The pathway is narrow, so there is not much room.

Method Peg out a straight line AB as long as possible. Peg out the next leg CD. Find the angle between the two legs by pegging out triangle CEF and measuring. The triangle could be CGH if it was more convenient. Continue to peg out legs and measure triangles to give the angles.

The Correction of a Survey Traverse

Surveys are never exactly right, but they can be corrected (see Fig. 15.14). A surveyor's sketch consists of a single line survey like the cliff path (see

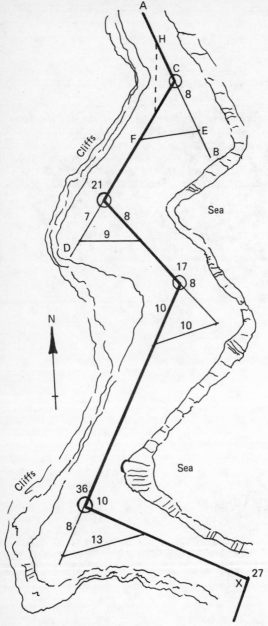

Fig. 15.13. A cliff path

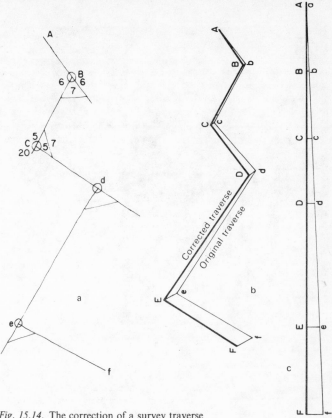

Fig. 15.14. The correction of a survey traverse

Fig. 15.14(*a*)). Fig. 15.14(*b*) shows the sketch drawn accurately according to the figures (the thin line) but point f should be at F. We know this from the Ordnance Survey map, which is accurate. The problem then is to alter our survey so that f moves to F **and all the other points move in proportion**.

The thin line in Fig. 15.14(*c*) shows the path as if it has been pulled out straight. Point f has been moved to F (the size of the error). AF is lined in heavily. Lines bB, cC, etc. have been drawn parallel to fF. They make a set of thin similar triangles. We can now correct the actual path (see Fig. 15.14(*b*)). Draw lines from b, c, etc. parallel to fF. Transfer length bB in this way from Fig. 15.14(*c*) to Fig. 15.14(*b*) and draw AB. Continue to transfer lengths until the traverse is corrected. The mistake fF will have been correctly distributed.

Fig. 15.15. A two bedroom bungalow built by Wates

By permission of Wates Built Homes Ltd

A modern two bedroom bungalow built by Wates is shown in Fig. 15.15. You will see how the central hall is placed to give access to the whole house; the corridor length is kept very short (the size of five doors); the kitchen has a double aspect, overlooking the front path and the garden and both living room and kitchen open directly into the back garden.

The drawing has been completed with a scribing machine (See the NC-scriber in Chapter 19 on Computer Drawing – Fig. 19.16.) The beds, sink, lettering, etc. are typical computer-drawn scriber symbols.

Fig. 15.16 shows a surveyor's sketch of the bungalow living/dining area. All measurements are entered as the surveyor walked. The numbering of each wall starts at 0. As the surveyor turns a corner, he starts again at 0. Measurements are placed at right angles to the wall. Doors are shown opening from the hinges.

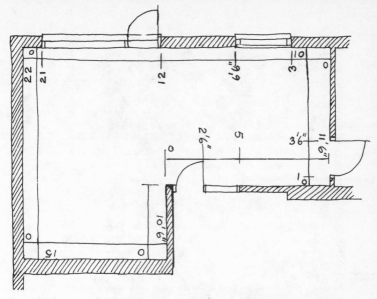

Fig. 15.16. A surveyor's sketch of the bungalow living/dining area of Fig. 15.15

A Surveyor's Sketch Plan of the Ground Floor of Two Seventeenth-Century Cottages Converted into One

Fig. 15.17 looks more complicated than it is. Think of it as one set of outside measurements and two separate inside sets. The plan should be read as the surveyor walked.

Rules for a Survey

Outside Start from one corner of a building and measure one side from 0. Write each measurement at right angles to the chain line. At the corner, start again from 0. A tape (15 m – 50 ft or 30 m – 100 ft) is better than a chain for measuring houses.

Inside Measure each wall in turn. When you come to a partition, start again at 0. Always measure a diagonal as few buildings have right-angled corners. Measure wall thicknesses at doors and windows.

The Finished Drawing

Decide on a suitable scale. The main problem is to get the angles of the building correct. Fig. 15.18 shows the start of the finished drawing being made from Fig. 15.17

Method Start at A. Draw the wall to scale 11.09 m long. Strike arcs to give point C. Find point B. Draw BX 4.92 m long. Strike arcs to give point C.

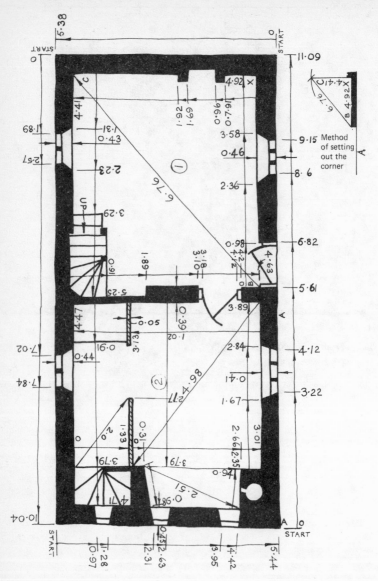

Fig. 15.17. The ground floor of two seventeenth century cottages converted into one

All measurements in metres

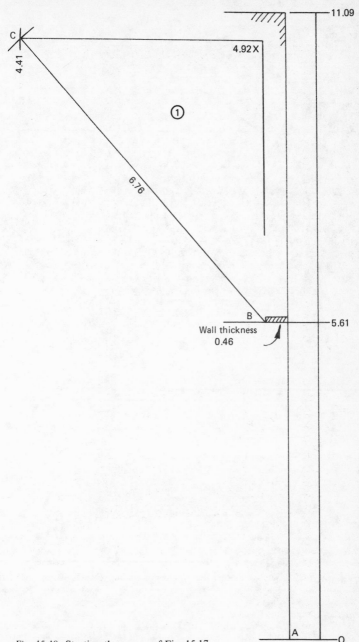

Fig. 15.18. Starting the survey of Fig. 15.17

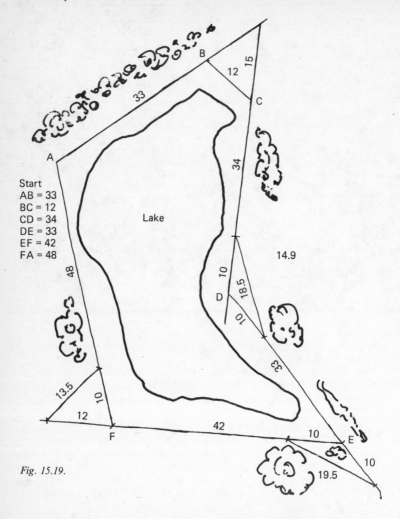

Fig. 15.19.

Start
AB = 33
BC = 12
CD = 34
DE = 33
EF = 42
FA = 48

Then continue the drawing. Notice that it has not been possible to measure the thickness of the end wall. It will have to be guessed. Remember that you will need correct corner angles in room 2 as well.

Questions

1 Set out the field shape for Fig. 15.4 to the given sizes. Draw the hedge DF from Fig. 15.6 (scale 2 mm = 1 m).
2 Draw the rough hedge round the rest of the shape. Set out the offsets and draw a surveyor's notebook entry for the hedges FB and EA on your drawing.
3 Draw survey 15.9.

4 Draw survey 15.12. Add building E which has one corner 27 m from C and 32 m from D. How far is this corner from the corner of A?

5 Draw the cliff path AX in Fig. 15.13. Point X is 12 m east of the correct position. Distribute the error and redraw the path beside the first.

6 Make a surveyor's sketch plan (like Fig. 15.16) of an actual room. An imaginary one is useless.

7 Make an accurate drawing (like Fig. 15.15) from your sketch to a suitable scale. Draw the scale. Add furniture and fittings to the room.

8 Make an accurate drawing of the ground floor of the seventeenth-century cottages in Fig. 15.17. Scale 1 : 50 (2 cm = 1 metre). Draw a scale.

9 The path surrounding the lake in Fig. 15.19 has been drawn by pegging out and measuring triangles at the corners. Make a finished survey.

10 Distribute any error and redraw the survey with a heavy line.

16
Electrical and Electronic Drawing

Many Design and Technology and Graphic Communication examinations now contain electrical and electronic drawings. (Electronics can usually be thought of as low voltage electrics.) This chapter is about the drawing of circuits; it will not teach you about the theory of electricity. For that you will need a specialist text such as *Electricity Made Simple*.

Fig. 16.1 shows a burglar alarm circuit as a picture drawing. It might be used to show the householder the sort of system that was going to be fitted. It might have the following description:

A mat outside the flat door is spring-loaded upwards. When a weight (the burglar) is put on it, it completes a circuit and rings a bell. The transformer takes the place of a battery, to give power for the circuit.

An electrician, however, does not need a picture. He needs a layout which will allow him to fit a burglar alarm to any house, so he needs a **circuit diagram**.

Fig. 16.2 is a circuit diagram, using standard symbols, for the alarm shown in Fig. 16.1. The components have also been named on this diagram but this is not normally done. Compare the two diagrams and check that they are electrically the same, with the same components in the same order. A circuit diagram shows the components and the wires connecting them, but not their actual positions. Two components which are 1 cm apart on the diagram, may be 50 m apart in fact.

Each component has its own agreed symbol. Fig. 16.3 shows a selection of symbols from the British Standards publication BS 3939. The complete volume is very expensive. You will need the booklet PD 7307: 1982 or a later edition when it comes out, which is a selection of 'Graphical Symbols for use in Schools and Colleges'. It also includes engineering drawing symbols, logos and many other things. The standard symbols should be learnt.

Examiners usually present you with a picture drawing like Fig. 16.1 showing the layout and ask you to draw a circuit diagram using the correct symbols. Sometimes they describe the layout in words.

Example 1 Draw a bell circuit using a transformer, a bell and a bell push. This is the simplest of all questions. (A bell push is a switch which is normally open.) Your drawing should look like Fig. 16.4. Check the symbols used against the list in Fig. 16.3.

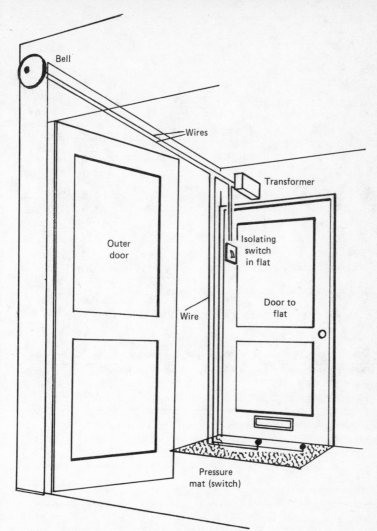

Fig. 16.1. A picture drawing of a burglar alarm circuit

Example 2 Draw a fused circuit showing a transformer and two lamps wired in parallel:

(*a*) separately switched
(*b*) both controlled by one switch
(*c*) separately controlled and also both controlled by a third switch.

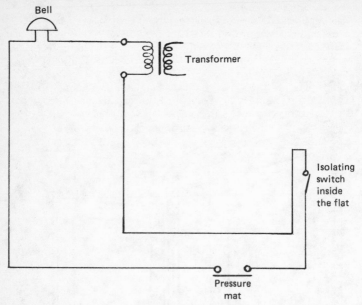

Fig. 16.2. A circuit diagram for the burglar alarm in Fig. 16.1

Description	Symbol	Description	Symbol
Direct current or steady voltage	—	Earth	⏚
Alternating	∼		
Positive polarity	+	Fuse	⎓
Negative polarity	−		
Primary or secondary cell	⊣⊢	Fixed resistor	▭
Battery of primary or secondary cells	⊣│⊢│⊢	Variable resistor	▱
Alternative symbol	⊣⊦---⊦⊢	Inductor + core	⏛
Example: 50 V battery	⊣⊦---⊦⊢ 50 V	Transformer	⏛

Fig. 16.3. Selected electrical and electronic graphical symbols from BS 3939.
Source: PD 7307: 1982

Fig. 16.5(*a*), (*b*) and (*c*) show a single fuse which protects the whole circuit. You could have fuses to protect each lamp separately. In this question, the examiner does expect you to know the difference between wiring in series and wiring in parallel. Lamps in **series** have the electricity running through the first lamp, then through another lamp and back to the battery. If one lamp burns out, this makes a gap in the circuit and the other lamp cannot stay alight. Christmas trees are often wired this way. Lamps wired in **parallel** are connected separately to the supply, and return lines to the battery. If one goes out, the others can stay alight.

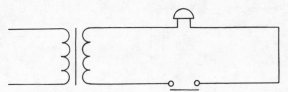

Fig. 16.4. A simple bell circuit

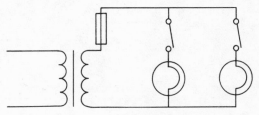

Fig. 16.5. (*a*) Two lamps separately switched

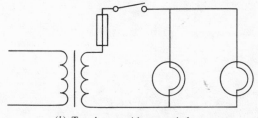

(*b*) Two lamps with one switch

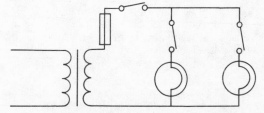

(*c*) Two lamps with separate switches and a common switch

Draw the three circuits shown in Fig. 16.5(*a*), (*b*) and (*c*). Draw a fourth one with a fuse protecting each lamp separately.

Example 3 In Fig. 16.6(*a*), two stage lights, separately switched, are controlled by a single dimmer. (A dimmer is a variable resistance which can reduce or increase the supply of current to the lamp and make it dim or brighten.)

In Fig. 16.6(*b*) two stage lights controlled by a dimmer are each separately fused.

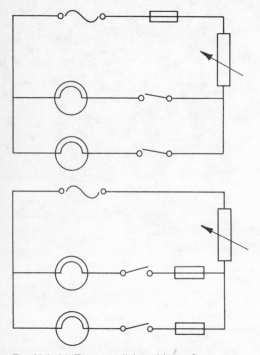

Fig. 16.6. (*a*) Two stage lights with one fuse
(*b*) Two stage lights each separately fused and switched

Example 4 The parking light circuit for a motor car is shown in Fig. 16.7(*a*). All the components are earthed to the frame of the car so that the return circuit is through the frame. (This arrangement saves wiring.) Draw the circuit. Note that a two-way switch is one that can be put in three positions. Two of them activate a circuit and the central position disconnects both circuits.

Fig. 16.7(*b*) shows the wiring diagram.

Now go back to Figs 16.1 and 16.2 and consider the following questions which give more design problems than mere wiring.

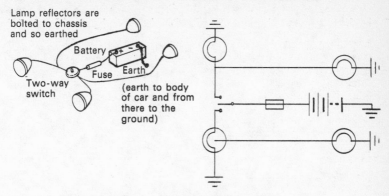

Lamp reflectors are
bolted to chassis
and so earthed

Battery

Fuse   Earth

Two-way
switch

(earth to body
of car and from
there to the
ground)

Fig. 16.7. (*a*) A picture drawing of a parking light circuit
(*b*) A wiring diagram of a parking light circuit

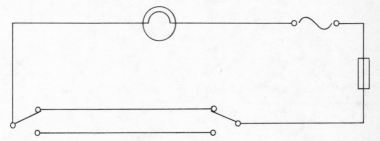

Fig. 16.8. A two-way switch circuit for upstairs/downstairs control of a hall light

1 if the burglar steps off the mat, the bell stops ringing. How could you
  wire it to give a continuous ringing tone?
2 How could one prevent the burglar from breaking the glass, putting his
  arm through, and switching off the alarm?

Possible Answers (see below)

Answers (These are only suggestions. There may be lots of others.)
1 Let it trip a switch which stays closed until reset.
2 Place any wiring inside the flat beyond arm's reach.
3 Put in a special switch which needs a key (perhaps the door key).

Two-way Switches

Fig. 16.8 shows the arrangement often used for lighting halls and stairways.
The lights can be switched on or off from upstairs and downstairs.

The Wiring Diagram for an Electric Toaster

In Fig. 16.9 Neutral and live lines are both switched. When the toaster is
heated, the thermostat breaks the contact only to remake it when the toaster cools. The body of the toaster is earthed.

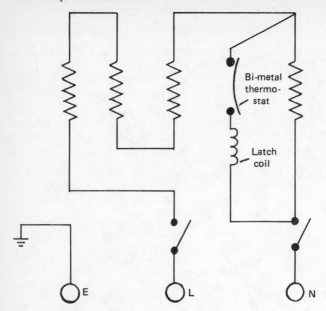

Fig. 16.9. Wiring diagram of a toaster
Courtesy of Murphy Richards

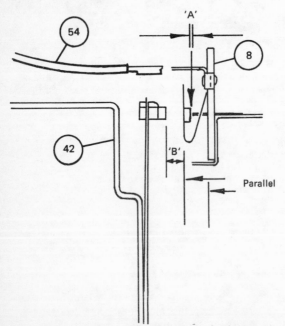

Fig. 16.10. A maintenance manual diagram of a thermostat
Courtesy of Murphy Richards

A Maintenance Manual Diagram of a Thermostat

In Fig. 16.10 the important details are the gap sizes 'A' and 'B'. This problem concerns mechanical engineering, not electrical engineering.

Question

Make a sketch like Fig. 16.1 to show the lighting of a hall.

17
Sketching

Some sketches should be thought of as pictures. They look more or less like the object they portray; are shaded or coloured; and can be full of feeling. Some sketches are attractive, some repulsive. They convey emotion. If they do not convey emotion, they are dead.

Other drawings are careful descriptions of facts. Leonardo da Vinci's dissections of the human body were a new way of looking at reality. They were careful, 'scientific' drawings of real corpses. They explained how the bones, tendons and muscles functioned. As far as possible, feeling was removed.

Fig. 17.1 shows a page from Constable's tiny sketchbook, reproduced at about full size. His large paintings were built up from dozens of small sketches like this to convey the romantic feeling of the pre-industrial English countryside.

Fig. 17.1. A page from Constable's sketch book

Fig. 17.2 shows a page from a book on Fortifications, by Daniel Specklin, 1589. This sketch of drawbridge mechanisms is what I have called a 'scientific' drawing. The book was sold all over Europe. The drawbridge has been raised by the worm and pinion gear (G) with the locking bar (M) pushed forward on the rollers (K) by lever (H). The bridge is also shown in the lowered position with the counterweight (C) clearly marked. The gearing (G) would not have been strong enough to lift the dead weight of the bridge without a counterweight. The drawing is full of facts. No doubt Specklin collected ideas for this engraving from many sources. Possibly nothing but the drawing was original, but it explained things clearly and this is why the book sold.

There is still a third type of sketch. This explores possible ways of solving a problem. The designer 'thinks' with a pencil, drawing possible solutions, discarding those that will not work, or are expensive, or difficult to make, and gradually improving the others. These sketches are methods of thinking. Nasmyth's Steam Hammer drawings are of this sort.

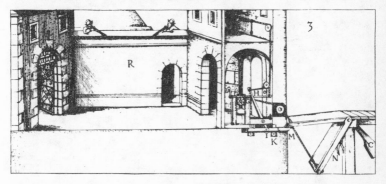

Fig. 17.2. Drawbridge mechanism, Daniel Specklin, 1589, Architectura, Bon Bestungen

James Nasmyth (1808–90), the famous engineer, was consulted about forging the immense wrought iron paddle shaft for Brunel's new ship, the *Great Britain*. The shaft was too big for any existing hammer. Sketch A in Fig. 17.3 shows a traditional tilt hammer. A water wheel turns a shaft with three cams which each lift the hammer in turn and suddenly let it drop. This device was simple, well tested over centuries and fast. It worked well for small articles but the big paddle shaft would 'gag' it, giving the hammer no room to fall. Nasmyth wrote:

> The obvious remedy was to contrive some method by which a ponderous block of iron should be lifted to a sufficient height above the object on which it was desired to strike the blow, and then to let the block fall down upon the forging, guiding it in its descent by such simple means as should give the required precision in the percussive action of the falling mass. Following up this idea, I got out my 'Scheme Book', on the pages of which I generally *thought out*, with the aid of pen and pencil, such mechanical adaptations as I had conceived in my mind, and was thereby

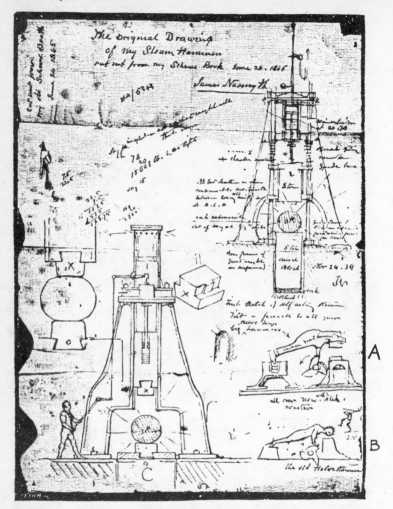

Fig. 17.3. First drawing of Steam Hammer, 24 November 1839, by James Nasmyth

enabled to render them visible. I then rapidly sketched out my Steam Hammer, having it all clearly before me in my mind's eye.

In little more than half an hour after receiving Mr Humphries' letter narrating his unlooked-for difficulty, I had the whole contrivance before me in a page of my Scheme Book. The date of this first drawing was 24th November, 1839.

This extract is taken from James Nasmyth's *Autobiography*, edited by Samuel Smiles 1883. The whole story is worth reading.

In Fig. 17.3 Sketch A shows a traditional tilt hammer with a small mouth. Sketch C shows the new hammer which was raised by steam. Steam was forced into the cylinder below the piston. This lifted the hammer. The operator opened the valve, allowing the steam to escape and the hammer to fall. The faster the valve was opened and closed, the lighter and more frequent the blows. A skilled operator could crack an egg in an eggcup, or flatten a billet of steel.

Nasmyth's account describes perfectly an engineer getting an idea and working it out in rapid sketches. Look at the detail of the dovetailing of the steel hammer head X into the huge cast iron weight. The central locking key is to prevent sideways movement. (The gib that pushes the dovetail across and the key into its slot, has not been shown.) In the middle of thinking about a new principle, Nasmyth was also bearing in mind fine details. This is the test of a good engineer.

An engineer must be able to use a pencil to clarify his own ideas and to explain them to others. Freehand sketching is very important.

Typical Workshop Sketches

Drawings like Fig. 17.4 need a lot of practice. Make small drawings of objects around you or in a workshop. Search for the cylinders, cones, prisms which make up the forms and draw as a series of blocks. Draw along centre lines. Keep all shading simple and bold.

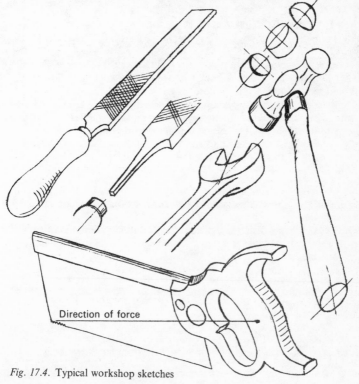

Direction of force

Fig. 17.4. Typical workshop sketches

These sketches show the objects, but they also show how the tools work. The file tang is a tapered square so that it will tighten as it is driven into the handle. The wooden handle may split so it has a metal ferrule round it.

The saw handle has been developed over generations to give a satisfactory, comfortable grip, with the forces pushing in the right direction.

The spanner mouth is at 30° so that the spanner can be reversed on a nut, in awkward corners. Tools have their own beauty but this is almost entirely because they are functional and unnecessary weight has been pared away.

Design Drawing

If we wish to design something new or to modify an existing drawing, we need to ask ourselves a number of questions.

Most drawings are of known things. A pillar box is a pillar box. It can have many shapes but it is a box for putting letters in. The letters are collected and delivered. If they are not delivered, there is no point in having a box. This is so obvious that it sounds silly, but it is most important.

The first question must always be, 'What does it do?' not 'What does it look like?' When you are asked to design something, ask yourself the following questions:

1 Why is it needed? What must it do?
2 Are there others in existence? If so, what are they like? How do they work?
3 Were they designed to do the same job? What are the differences between what they do and what the new design must do?
4 Would new materials change the design?
5 Are there new methods of manufacture which would change the design?
6 Are the existing ideas in the present design still in patent. (Most patents last about 16 years and must not be copied during that time. Also, ideas which have been patented and run out, cannot be re-patented, so they must not be part of your patent application but you can, or course, use them.)

Designing movements or actions for machines is different from designing known objects. You are thinking of actions, not the finished article. Let us take an example.

The **Pedal Bin** is a common article today but it is quite modern. Imagine that a pedal bin had never been invented. The process of design might have gone like this:

Design Brief It is important not to get your hands dirty when cooking. Design a container for kitchen waste with a lid which can be opened without getting your hands dirty.

Ideas There are no such bins in existence because we are going to invent the very first one. All the bins we can find are without lids, or the lids are picked up by hand. This is very unhygienic. We must start afresh.

1 A plastic bag – how can it be opened? Put a spring on it. How?

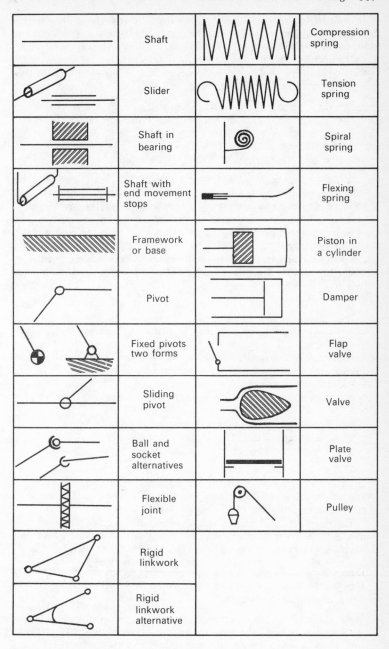

	Shaft		Compression spring
	Slider		Tension spring
	Shaft in bearing		Spiral spring
	Shaft with end movement stops		Flexing spring
	Framework or base		Piston in a cylinder
	Pivot		Damper
	Fixed pivots two forms		Flap valve
	Sliding pivot		Valve
	Ball and socket alternatives		Plate valve
	Flexible joint		Pulley
	Rigid linkwork		
	Rigid linkwork alternative		

Fig. 17.5. Symbols which describe mechanical functions

2 A wooden box. It can have a lid – hinged? removeable? – How – No hands – worked by electricity? – dangerous near a sink and water – also expensive. Open with foot? – How? – Dangerous with both hands holding rubbish? A pedal? – You still have to take one foot off the ground. – Can we keep the heel on the ground and push down with the toe? That way we can keep our balance.

So far the ideas have been in words, now the designer needs some way of showing **functions** – the way things move. There is no British Standard for these at present but the shapes in Fig. 17.5 are often used. They are very simple to understand.

Fig. 17.6 shows the designer 'thinking' with a pencil.

(*a*) Shows the amount of movement of the pedal and the amount of movement necessary for the lid.

(*b*) Shows a possible lever system. The movement of the lid is far too small.

(*c*) Shows the pivot point moved forward, but the movement of the lid is still too small.

(*d*) Shows that the hinge point of the lid has been moved back, outside the edge of the bin. This has increased the lid movement and the bin will now work.

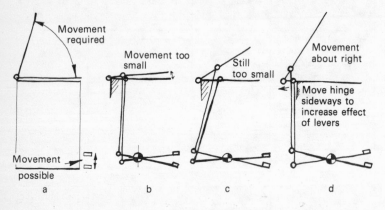

Fig. 17.6. Stages in planning the movement of the pedal bin

Fig. 17.7 shows the finished drawing of the pedal bin. Having designed it, you have a new idea. What would happen if the pedal revolved like the handle of a door when you pressed down on the end? Perhaps it would be better. Back to the drawing board and design it.

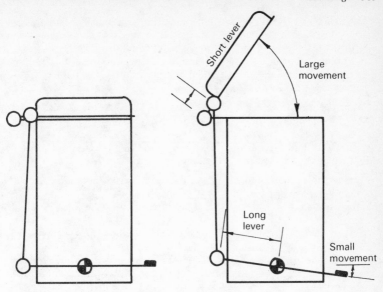

Fig. 17.7. Pedal waste bin

Sir Christopher Wren's Original Drawing for St Paul's Cathedral (Fig. 17.8)

An old rhyme goes:

Sir Christopher Wren,
Went to dine with some men,
He said, 'If anyone calls,
Say, I'm designing St Paul's'.

If that rhyme were true, Fig. 17.8 might be the sketch he drew on the coffee house menu. Again, he was 'thinking with a pencil'. The centre of this sketch is the parabola which holds up the dome. Nobody was to see the dome for generations. Probably Wren himself did not know exactly what it would look like for years, but the parabola is there from the start.

Do not restrict your sketching to engineering parts. Sketch buildings, flowers (see Figs 17.9 and 17.10) and anything else. Good architects and engineers have a sketchbook in their hands all the time. These sketchbooks are worth finding. Look at Charles Rennie Mackintosh's drawings of flowers. He was a famous architect but his flower paintings are outstanding. Good sketching requires good observation and that is vital in all walks of life.

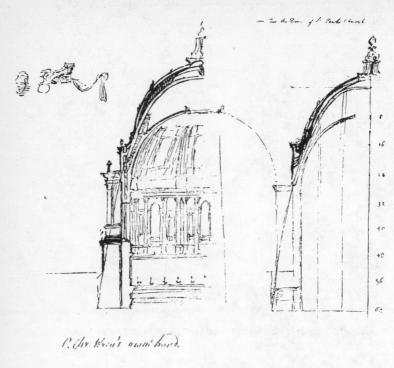

The Dome of St. Paul's Cathedral. From a Sketch by Sir Christopher Wren. (British Museum.)

Fig. 17.8. The dome of St Paul's Cathedral. From a Sketch by Sir Christopher Wren.

Copyright British Architectural Library/RIBA

Exercise

PRACTICE WORK

Draw any of the following objects to show their forms and functions:

> lap dovetail joint
> screwdriver
> three-jaw chuck
> blacksmith's forge and hood
> wheelbarrow.

Fig. 17.9. Typical small holiday sketches

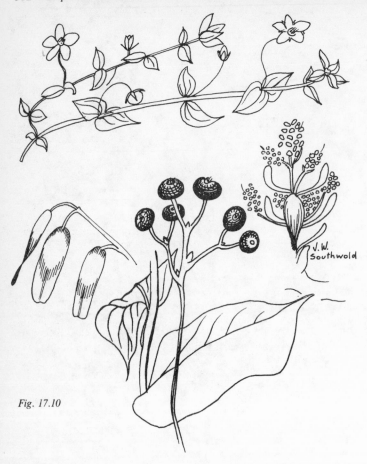

Fig. 17.10

You will think of dozens more. Where possible, draw them from real life, as simply as possible. They should not take more than five minutes each. Then draw another version so that you finish up with several sketches of one article on the same page.

Instructional Drawings

Make a drawing of an article for a catalogue and number the parts.

Examples: a food mixer, a car battery, a stapler.

Drawings for Enjoyment and Understanding

Draw some real flowers, berries, etc. Small weeds and twigs very good to draw because they are so continually surprising. Their forms are endless.

18
Logos and Pictograms

Logos

More and more firms and organisations have symbols as trademarks on letter heads. They are often more easily remembered and understood than names. Gas companies have flames; travel companies have aeroplanes or globes; caring organisations have protecting arms.

Heraldic Banners were logos for troops, who could not read or who were foreign mercenaries who spoke a different language to rally around. Again, we find profiles used in heraldry because they were the simplest shapes to recognise. The fine beast, in Fig. 18.1, with the body of a lizard, the wings of a bat, the feet and legs of a cockerel, and the head of a wolf, knows how to arrange himself on the page.

Fig. 18.1.

Logos must all be simple, easily recognisable, and make an impact. A logo which is difficult to understand is useless. All logos are derived from real things. Egyptian hieroglyphics were drawings of objects. Later, some of them became abstract words or even letters, but they started from doors and snakes and other real things.

Egyptian Drawings

A number of simple Egyptian hieroglyphics which were derived from architectural details are shown in Fig. 18.2.

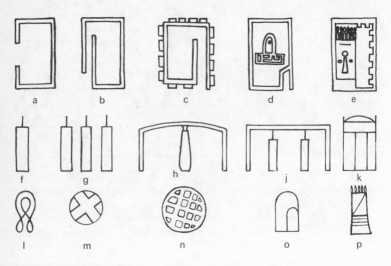

Fig. 18.2. Egyptian hieroglyphics derived from architectural details

Archaic Period

(*a*) A plan of a house.
(*b*) A courtyard.
(*c*) A fortified courtyard.
(*d*) A castle entrance inside the outer wall.
(*e*) A closed courtyard. The right half shows the bastions on the inside of the enclosing wall. This half is a plan. Inside, on the left, is the elevation of the palace tower.

The Middle Kingdom

(*f*) A pillar with a top tenon.
(*g*) A pillared hall.
(*h*) A vaulted building with a central row of pillars.
(*j*) A vaulted shrine with four corner posts. All four posts are shown.

Fig. 18.3.

This was usual in Egyptian drawing. Their rules of drawing were very rigid: artists were not free to draw what they pleased, but had to obey the custom. Everything had to be shown complete and at its widest. The eye was seen from the front; the nose in profile; the shoulders at full width. Every finger had to be spread out, even if this made it difficult to hold a pen. The four posts shown here (*j*), are really one at each corner but nothing could be hidden so all four had to be shown.

(*l*) The looped rope used by surveyors to mark out the fields afresh after the annual flooding. This is not shown as a right angled triangle but looped over the assistant's shoulder (see Fig. 15.2).

(*m*) A village.

(*n*) City (Archaic Period).

(*o*) A vault showing the elevation of the front and a perspective view of the end.

(*p*) The elevation of a battered (tapered) palace tower with a kheker freize at the top and horizontal stripes at the base. The rectangle at the top, with the diagonal line, represents the same building in plan.

Fig. 18.3 has been redrawn in black and white from a small coloured casket found in the tomb of Tutankhamun. The black and white picture reveals that everything is in profile or else in full face. Each animal is twisted and distorted in death. Picasso used the same construction in his picture 'Guernica', depicting the destruction of the Basque capital by Franco during the Spanish Civil War. In Picasso's picture a terrified horse rears and writhes in agony. The Egyptian lions and the Spanish horse are logos of ruthless destruction. Crucifixion pictures show the same pathetic twist of the body and evoke a feeling of extreme pity.

Other examples of logos are given below.

Fig. 18.8 shows interlinking initials. This logo was designed by the Omega Workshop (Roger Fry, Duncan Grant, Vanessa Bell *et al.*) some time before 1919 and painted on the backs of a set of chairs for the novelist Virginia Woolf. This was years before Volkswagen was ever heard of.

Fig. 18.9 shows a delightful notepaper design from Ultimate of Halifax, who specialise in mountaineering equipment. The original is in four tones

Pictorial marking of goods in transit

| This way up | Fragile Handle with care | Keep dry | Keep away from heat |

Fig. 18.4. Pictorial markings for goods in transit
By courtesy of British Standards

Fig. 18.5. A heart-pacer warning beside an electronic barrier in a public library

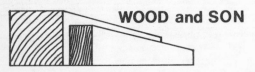

Fig. 18.6. Some logos are puns

Fig. 18.7. Colour reversals are effective. Drawn by Tom, aged eleven.

Fig. 18.8. Interlinking initials

Fig. 18.9. A notepaper design for a firm making mountaineering equipment
By permission of Ultimate Equipment Ltd

of blue which give an effect of perspective. For this book, the blue tones have been replaced by architectural tints in four tones (see Chapter 20).

Logos for the Olympics are essential if people, speaking all the languages of the world, are to be at the different starting points at the correct times. The sports symbols shown in Fig. 18.10 have taken typical sporting actions and stylised them. All lines are vertical, horizontal, or at 45°. Strangely enough, this has not limited the artist, but set him free to draw wittily. It also makes the drawings into a set. You 'read on', wondering how the next problem will be solved.

Fig. 18.10. Sports symbols taken from a Letraset dry-transfer sheet
By permission of Letraset Ltd

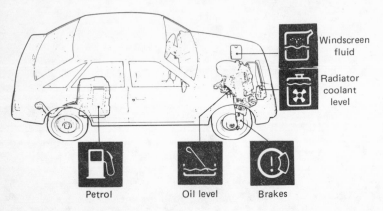

Fig. 18.11. Ford's logos
By permission of Fords Great Britain

Logos used by Ford's for their Cars

Ford have used logos to show the locations of the petrol, coolant, etc. simply and in an international sign language (see Fig. 18.11). The latter is very important for any firm selling to countries all over the world. The signs avoid the dangers of bad translation. Incidentally, Fig. 18.11 is an excellent example of 'glass body' drawing. The outside is drawn but the inside is fully exposed.

Women in Engineering

Fig. 18.12 is a very good logo developed for the magazine *View* by the Central Office of Information. The hard hat stands for Civil Engineering and the punch tape for modern Production Engineering. The label is bold and the whole logo very attractive.

Fig. 18.12. Engineering recruiting logo

Published by permission of the Department of Industry

Municipal Heraldry

Cities, boroughs, councils, have had their own crests (which are in fact logos) for centuries in some cases and it is interesting to trace the changes in design. We can only touch on it, but the subject could bear close study.

The City of Nottingham is old and established, as its heraldic crest shows

Fig. 18.13. Crest of the City of Nottingham
By permission of the Chief Executive

Fig. 18.14. Logo of Nottinghamshire County Council
By permission of the Clerk of the Council

(see Fig. 18.13). The symbols of stags, a castle, a shield with coronets, are those of the nobility. Even the stags have coronets. The whole design is aristocratic and proud. Modern heraldry is much simpler and more earthy, related to the area instead of power.

Nottinghamshire County Council, a democratic foundation, uses a simple outline letter 'n' with an oak tree, in an attractive outline pattern of leaves and acorns, inside it (see Fig. 18.14). The eight acorns represent the number of District Councils within the County. The tree is the Major Oak, part of Sherwood Forest, the very symbol of Nottinghamshire.

The Borough of Torfaen has the effective logo shown in Fig. 18.15. It combines the wavy lines representing the Afon Lwyd (the Grey River), which runs the length of Gwent's eastern valley, and the initial 'T'. It is simple, direct and attractive.

Halton is a new borough, made by combining Runcorn and Widnes. The logo of the Borough of Halton is shown in Fig. 18.16. The castle symbol on the lower right is Halton Castle which is situated on the top of Halton Hill – a Norman castle, now in ruins, that dominates the skyline at Runcorn. The factory buildings – top left on the logo – represent the industrial skyline of Widnes. The two are separated by the river Mersey – the broken lines – and are linked by the symbolic representation of the high-level Runcorn–Widnes Bridge which is the most dominant physical feature in the locality.

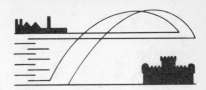

Fig. 18.15. Logo of the Borough of Torfaen
By permission of the Public Relations Officer

Fig. 18.16. Logo of the Borough of Halton
By permission of the Chief Executive

Finally, logos are widely used in business. Fox Talbot was a pioneer photographer so it was natural for a photographic firm to use the name. Their logo is a neat pun on the name 'Fox' but the logo is also a visual pun (see Fig. 18.17). The two legs and tail of the fox are a pun on the tripod, something that would not have been possible if a man had been standing behind the camera.

This logo combines many of the important points about a good logo. It is simple, elegant, witty and makes an immediate impact, without taking itself too seriously.

Fig. 18.17. The Fox Talbot logo
By permission of Shadow Photographic Ltd

Pictograms

Bar Charts

We are all familiar today with graphs and bar charts (see Figs 18.18–
18.22). Fig. 18.18 shows monthly totals as a series of columns. Fig. 18.19
has turned them into skyscrapers. Fig. 18.20 shows two sets of figures
(Home Sales and Foreign Sales) on each bar. Bar charts can be drawn in
perspective (see Fig. 18.21). Any vanishing point can be chosen. 18.22 is a
scale drawing of the distances of the planets from the Sun. It is a sort of
bar chart on its side.

   Bar charts can be drawn in many ways. Piles of coins, columns of oil
drums, traffic jams of cars, sacks of coffee, will all make bar charts. Their
heights and distances apart are important. The rest is decoration.

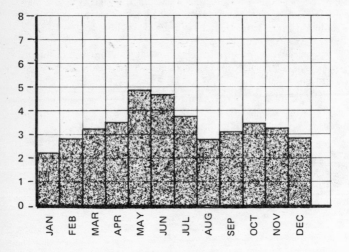

Fig. 18.18. A monthly bar chart
Totals in thousands of pounds

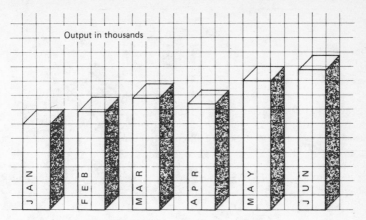

Fig. 18.19. A skyscraper chart

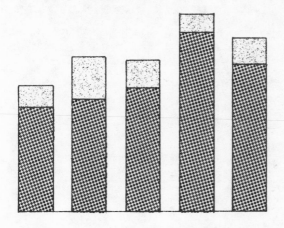

  Foreign sales

  Home sales

Fig. 18.20. A five-year comparative bar chart

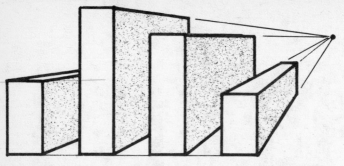

Fig. 18.21. Bar chart in perspective

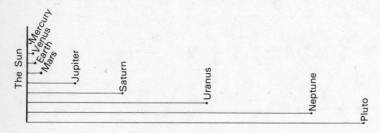

Fig. 18.22. A diagram showing the distances of the planets from the Sun

Pie Charts

Politicians talk of 'dividing up the nation's cake'. They mean that every year, a nation, or a firm, or a family, has a certain amount of income which must be shared. These divisions are often given in percentages. Special percentage pie charts can be bought. Fig. 18.23 is one example. Fig. 18.24 shows an elliptical pie chart, which is more decorative. They are both divided into 100 equal parts.

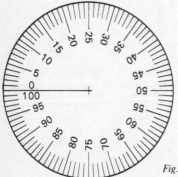

Fig. 18.23. A blank percentage pie chart.

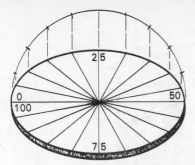

Fig. 18.24. The method of drawing an elliptical pie chart

Fig. 18.25 shows a list of sales in percentages. Fig. 18.26 shows the same figures as a pie chart. The proportions are much easier to compare than in the list. Removing a slice, as shown in Fig. 18.27, can add interest.

%	Sales in the past twelve months
47	Bicycles
21	Bicycle accessories
23	Electrical appliances
2	Electrical spares
7	Miscellaneous

Fig. 18.25.

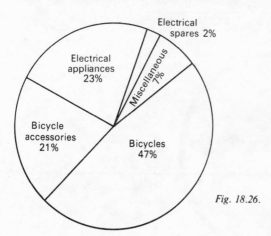

Fig. 18.26.

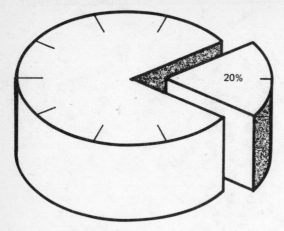

Fig. 18.27. A removed slice

Visual Aid

Graphs, bar charts and pie charts are often ingenious and even amusing. They are worth collecting and making into a display. Remember that it is easy to tell half truths or even downright lines by selecting figures which appear to support a particular argument, so look out for these as well.

Questions

1 Design three logos for a fête. They should be the same shape and size and be in the same style. Possible subjects might be – Pony Rides, Cake Stall, White Elephant Stall, Coconut Shies, etc.
2 Design three or more logos for a Children's Zoo showing different tame animals. (See Fig. 18.28 for example.)

Fig. 18.28.

Fig. 18.29. Logo for a lift

3 Collect as many book publishers' logos as possible. They call them colophons. The Penguin is possibly the most famous, but there are many (see page iii of this book for instance) and they are well worth drawing or photocopying. Never cut one out of a book.
4 Design a logo for a Model Engineering Exhibition. It can embrace model engineering as a whole or just one aspect of it.
5 Design logos for any three shops, showing their merchandise, not their names.
6 Design hotel logos for any of the following – showers, bathroom, lifts (example given – see Fig. 18.29) telephone, toilets (male/female), cashier.
7 Design three logos to be used in a public park for any of the following – first aid, drinking fountain, refreshments (tea/coffee/wine) cutlery, brass band.
8 Design a logo for a dance studio. Include the name of the studio in some way.
9 Design logos for three sports in the next Olympic Games.
10 Design a walkers' trail in a Nature Reserve showing routes to bird hides, wild flowers, etc. Design the actual trail in some interesting form, e.g. animal tracks.

19
Flow Charts

Flow charts are simple methods of explaining processes in a series of logical steps. One could make a flow chart of brewing a cup of tea or boiling an egg. Most flow charts, however, are more purposeful than this. They can show, for example, the machining steps in making a finished article from a blank (a roughly formed piece of suitable size); transferring goods by road, rail and aeroplane, to markets abroad; deducting a sum of money from your bank account and transferring it to the shopkeeper's account when you buy something; or drawing a picture by computer. Each of these is a process with logical steps and the flow chart can show them in order.

Flow charts are something like electrical circuits, with the current travelling in a continuous path from one side of the battery to the other. In a circuit, the electricity changes things (lights lamps, rings bells, etc.) or its own path may be changed (by relays, etc.). The circuit must be complete or it will not work. **Flow charts too, must be continuous.**

Flow charts are used for many purposes, but most of the symbols used are common to all. The most important symbols are shown in this book but you will need a copy of PD 7307: 1982 or a later edition if it becomes available, *Graphical symbols for use in schools and colleges*, British Standards Institution. This booklet also includes electrical, machine drawing and other symbols.

Fig. 19.1 shows the typical symbols. Always put the description of an action inside a symbol. The symbol may be made larger to contain the word(s) but, if the wording is very long, put a keyword in the symbol and the full explanation at the side.

Some processes depend on others. Thus, only when **both** actions are complete, can the process continue (see Fig. 19.2). Flow lines must be either:

(*a*) horizontal, left to right
(*b*) vertical, top to bottom

If you need to draw diagonally, take two strokes (see Fig. 19.3):

(*a*) Vertical and horizontal
(*b*) and (*c*) vertical and horizontal or horizontal and vertical

Never draw a diagonal line.

Fig. 19.5 shows the use of a **loop**. A loop is a repeat process. This chart

Terminal/interrupt

This symbol represents a
terminal point in a flow
chart, e.g. a start, stop,
halt, delay or interrupt.

Process

This symbol represents
any kind of processing
function, e.g. the process
of executing a defined
operation or group of
operations resulting in a
change in value, form or
location of information,
or in the determination of
which one of several flow
directions is to be followed.

Decision

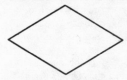

This symbol represents a
decision or switching type
operation that determines
which of a number of
alternative paths is to be
followed.

Preparation

Input/output

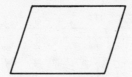

This symbol represents an ,
input/output function,
where the data medium
is not specified.

Connector

This symbol represents
an exit to or an entry
from another part of the
flow chart. The symbol is
often used to break a flow
line in order to avoid
cross-overs and long flow
lines or when a flow chart
takes up more than one page.

Annotation

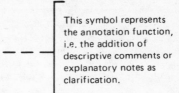

This symbol represents
the annotation function,
i.e. the addition of
descriptive comments or
explanatory notes as
clarification.

This symbol represents
modification of an
instruction or group of
instructions, e.g. setting a
switch, modifying an index
register, or initializing
a routine

Fig. 19.1. Symbols of data processing flow charts
Source: PD 7307: 1982

Parallel ↓ processes

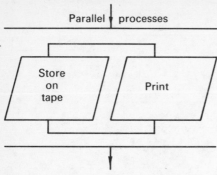

Fig. 19.2. When the material has been stored on tape *and* it has been printed the process can continue

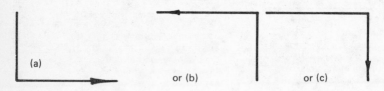

Fig. 19.3.

shows the machining process in which articles have to be tested for size. They are ground at 1 and tested at 2 and at 3. The machine has to make a decision. If the size is correct, the part is stored and the machine continues to grind more parts. If the part is too large, it is returned by the loop for regrinding. If the part is too small, it is scrap. The machine must be stopped before it makes any more scrap. The machine stops automatically and a sign lights up calling the tool-setter.

Fig. 19.6 shows the production of a new article and the consultation with Marketing (who should know what may and may not sell) at all stages.

Flow charts are also used to arrange processes in **time**. A building site, for example, must receive materials in reasonable amounts which will not clutter up the site, in the correct order, and at the right time so that work can proceed steadily. This might mean:

1 **Foundation laying** Time two weeks.
  Materials: sand, gravel, cement – to be delivered twice a week.
2 **Bricklaying** Time nine weeks.
  Materials: bricks, soft sand, cement – to be delivered three times a week.

This can be made into a flow chart (see Fig. 19.7) which is, in fact, a simple diary.

On any building project the different trades have to work in the correct order. Sometimes work depends on other work having already been done.

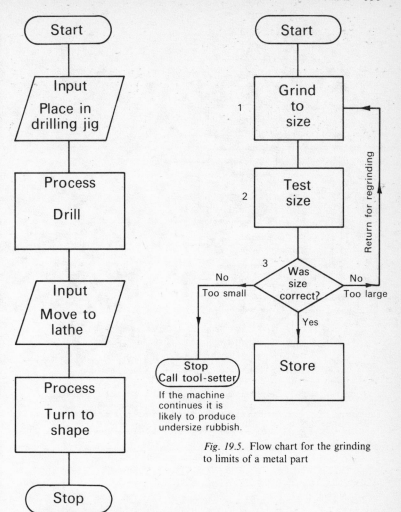

Fig. 19.5. Flow chart for the grinding to limits of a metal part

Fig. 19.4. Flow chart for the machining of a metal blank

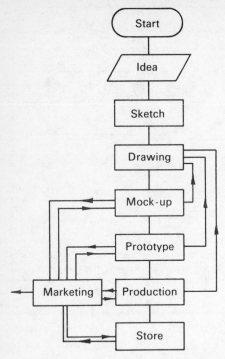

Fig. 19.6. Flow chart for the design and production of a new article

	Week 1	2	3	4	
M	Deliver s/g/c	s/g/c	b/s/c	b/s/c	
T					
W			b/s/c	b/s/c	
Th	s/g/c	s/g/c			
F			b/s/c	b/s/c	
S				Continue for	

Key   b = bricks
            c = cement   further 7 weeks
            g = gravel
            s = sand

            Weeks 1 + 2 – Foundation laying
            Weeks 3 – 11 – Bricklaying

Fig. 19.7. A delivery flow chart

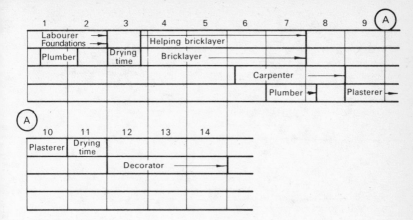

Fig. 19.8. A building flow chart

Sometimes trades can work at the same time as each other. All this can be laid out on paper (see Fig. 19.8). Each main process can be timed, and other trades – electricians, plumbers, etc. – warned when they are likely to be needed.

Critical Path Analysis (Network Analysis)

Where a project is a major undertaking, such as a large construction job, the multi-facet operation is controlled by a process known as critical path analysis. Other names for the same general approach are network analysis and Pert (programme evaluation and review technique). The problem is to devise the most economical method of working so that each part of the work is begun at the optimum moment and is ready for use when it is required in the overall plan. This will ensure that the work is approached in the most direct way so that the journey to final completion is as short as possible – made along the critical path.

Critical paths are arrived at by drawing a special type of diagram called a network. By convention a network is drawn from the left-hand side of the page, where the events and activities start, to the right-hand side of the page, where they finish. Each event is numbered, and each activity is indicated by a line with an arrowhead at the right-hand end where the activity finishes. Against each activity line the expected duration of that activity is shown. The general idea is best understood by looking at a simple network, such as [Fig. 19.9].

The critical path is that series of essential events which has the longest time-path; in other words, the project cannot be completed in less time than the critical path indicates. Thus events numbers 3, 4, 6, 7 and 8 are the ones that decide the critical path in Fig. 19.9. The project will take 22 days to complete. Activities on the critical path have to be carried out in the time stated. Activities which are not on the critical path can be allowed to 'float'; we can get activity 1–4 done at any time during the 10 days that 1–3 and 3–4 are being completed, so long as we start it by day 6 so that it gets done in time.

Clearly, critical path analysis is an art which has to be learned. The actual stages in planning a project may be listed as follows:

(*a*) State clearly what has to be done.
(*b*) Break this down into events, activities and duration times.

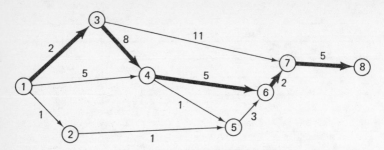

Fig. 19.9. A simple network showing the critical path (times are in days)

Notes:

(*i*) Each event is shown by a circle and they are numbered from left to right of the page, in the time sequence in which they occur.

(*ii*) An event is not reached until all the activities leading into it have been completed. The only event which has no activities leading into it is the starting event 1.

(*iii*) Activities are shown as arrows. They are clearly identifiable tasks to be performed, usually by one man or one team of men. The length of the arrow has no connection with the duration of the job, it only helps to join up the network. The duration of the activity is marked alongside the arrow in agreed units, like hours or days.

(*iv*) An activity cannot begin until the event it starts from (the tail event) has been 'reached'. (See (*ii*) above.)

(*v*) An activity is described by naming its tail event and its head event. Thus activity 1–2 takes 1 day and activity 1–3 takes 2 days.

(*c*) Draw the network.
(*d*) Analyse the network and decide how to schedule the activities.
(*e*) Check the schedule against the network.
(*f*) Institute progress controls over the project as it moves through to completion.

Reprinted by permission from *Secretarial Practice Made Simple* by Geoffrey Whitehead.

Making a Flow Chart

Think carefully about the subject first. Research it if you have time. Make small sketches of the separate stages in the flow. Join the stages logically, without overlapping. Use connectors if the chart becomes too long.

Check (completely separately)

1 That the flow 'works' and does what is required.
2 That you have used the correct symbol each time.

Questions

1 **Building a House** Set out a calendar of five working days a week. Make a flow chart showing when different trades will work and their lengths of work to complete their parts of the building. Allow for drying times and arrange a sensible order of work.

Trades

labourer	electrician	plumber
bricklayer	plasterer	slater
carpenter	decorator	scaffolder
gas-fitter		

Some will need to come in two or more times (e.g. the plumber will need to put in his mains supply and pipes, and then return to fit the wash basins etc.).

There is no correct solution to this problem. It is the sort of problem every builder faces and never manages to get quite right. The question has been included to make you realise how much information is required before one can come to a decision. How long will it take the labourer to dig the foundations? Who works first, the carpenter or the plasterer. It is really a subject for a class discussion.

2 Make a flow chart to show the flow of mail into an office. It is delivered by the postman, opened and sorted by whom? It is distributed by whom? How will the firm know later when the letter arrived? If the letter contains money or a cheque, how does the firm know? What happens to the cheques and money? Are they distributed with the letter? What level of person opens the letters? What checks are there on honesty?

3 Make a flow chart for replacing a washer on a tap.

4 Make a flow chart on replacing a fuse (remember safety precautions). Which colours are which. (Are you colour blind? Seven per cent of men are and 0.5 per cent of women. Should your flow chart point this out?)

5 A lecturer in a college has her class 'hand-outs' typed by a typing pool. The pool takes two days to type and return notes. She particularly needs good notes for a series of lectures on Wednesday evenings. The course lasts thirteen weeks. Draw a flow chart for the handing in, typing, returning and use of the notes required each week.

20
Computer Aided Design (CAD)

Computers are becoming more and more important in engineering design drawing, in producing display graphs, and for advertising. Complicated drawings need powerful computer memories and the ability to calculate thousands of positions rapidly. However, as computers become smaller, graphics are coming into our homes. Many students will have had some experience of this, even if it is only playing Star Wars and firing guns at monsters. The following example is a simple introduction to what techniques are available in computer graphics. It explains the principles involved. To use an actual machine, you will need the accompanying instruction book of course.

Drawing with Coordinates

Computers draw by instructing dots of light to form lines on a television screen. Each dot has an x and a y co-ordinate and is calculated separately. In Fig. 20.1 point A has co-ordinates xa and ya. Some systems count from the top left hand corner (Third Angle) and some from the bottom left (First Angle). In this book we start from 0 at the bottom left. Two points A and B can be positioned on the screen and the computer be instructed to join them. The line is not continuous, but a series of short parallel lines as near to a straight line as possible. Each line is a series of steep steps or shallow ones (see Fig. 20.2).

A series of points can be joined to form a graph with a base line (or axis) below (see Fig. 20.3). The graph can be tilted with the axis at say 30°

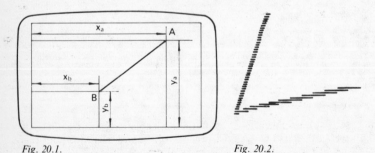

Fig. 20.1.   Fig. 20.2.

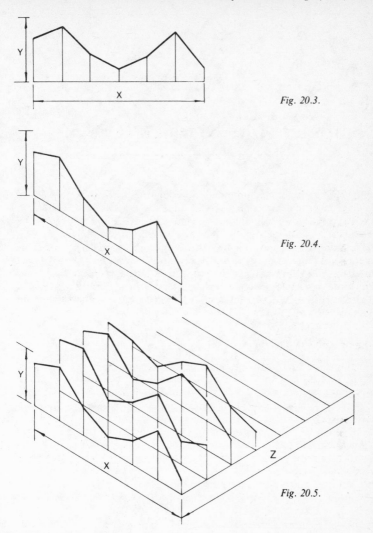

Fig. 20.3.

Fig. 20.4.

Fig. 20.5.

(see Fig. 20.4). A row of graphs can be placed one behind the other (see Fig. 20.5). Each point on the set of graphs has been calculated by finding **three** figures – the x, the y and the z co-ordinates.

Computer-drawn Maps

Fig. 20.6 is a three-dimensional map of a small area in North London. The points were calculated by laying a network of squares over a contour map as in Fig. 20.7 and calculating a series of graphs on the z axes.

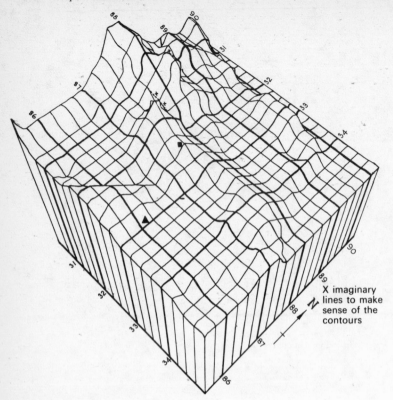

Fig. 20.6. Three-dimensional computer-drawn base map of Stoke Newington in North London

Line z = 5 would read:

x	y	z	x	y	z
1	35	5	11	20	5
2	30	5	12	15	5
3	30	5	13	20	5
4	25	5	14	20	5
5	20	5	15	20	5
6	20	5	16	20	5
7	20	5	17	20	5
8	20	5	18	15	5
9	25	5	19	15	5
10	20	5	20	15	5

The x number progresses from 1 across the page.
The y number rises and falls with the contour height.
The z number is constant for this line.

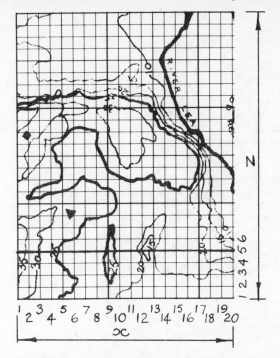

Fig. 20.7. A contour map with a square grid for calculating x, y and z co-ordinates. This is a larger area than is shown in Fig. 20.6.

Calculating the figures for this map involves three numbers for each point. A 20 × 20 square involves 20 × 20 × 3 = 1,200 numbers. When all the numbers have been collected, they have to be typed into the computer. This is a long, time-consuming process. The making of a map like that shown in Fig. 20.6 can cost several hundred pounds. There is a cheaper way which we will see later.

Once the map has been produced, it can, of course, be used in many ways. Figs 20.8 and 20.9 are examples.

Oil exploration firms use very powerful computers to transform vast amounts of geological data into cross sections and contour maps of this sort, for study by their geophysicists.

The Digital Tracer

There is a quicker way of putting the x, y and z co-ordinates into a computer, using the RD digital tracer.

The digital tracer (see Fig. 20.10) consists of a pivoted arm which carries a cursor (a disc with two crossed lines). As the arm moves, it feeds the position of the cross hair lines on the cursor into the computer. The slower it moves, the more points will be fed in. The cursor is moved along the **contour** this time. The y co-ordinate is the height of the contour. This

Fig. 20.8. A topographical sketch of the area over the base map. Notice how the names of the places begin to have meaning.

height is set on the computer and stays the same height until all contours at that height have been drawn. As the cursor moves along the contour, a series of x and z co-ordinates is fed into the computer but these also appear on the screen as a curve.

When one height of contour has been completed, the y co-ordinate (the contour height) is changed in the computer and the next contour is traced. This gradually builds up on the screen a picture which can be printed. Once the co-ordinates have been stored in the computer memory, it can be instructed to draw the map again from a different direction.

To Draw a Three-dimensional Contour Map using a Digital Tracer

1 Set the map at an angle of 45°. (See Fig. 20.11(*a*)).
2 Trace the highest contour, which will appear on the television display as Fig. 20.11(*b*).
3 **Either** instruct the computer to draw the next set of contours lower down, **or** move the map up by a small amount.
4 Trace the next lower contour line which will appear as shown with narrower gaps at the top than at the bottom. (See Fig. 20.11(*c*).)
5 Repeat 3.
6 Repeat 4. This will appear as Fig. 20.11(*d*). Delete any lines which are 'hidden' by the contour lines above, or, better still, do not draw any 'hidden' lines.
7 Repeat 3 and 4 until the map is complete.

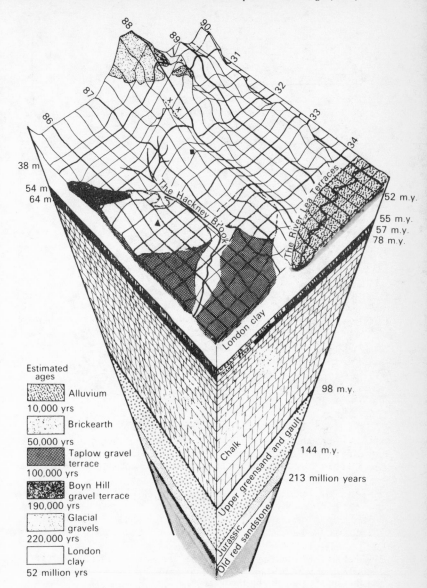

88
89
90
31
32
33
34

87
86

38 m
54 m
64 m

The Hackney Brook

The River Lea Terraces

52 m.y.
55 m.y.
57 m.y.
78 m.y.

London clay

Estimated
ages

Alluvium
10,000 yrs

Brickearth
50,000 yrs

Taplow gravel
terrace
100,000 yrs

Boyn Hill
gravel terrace
190,000 yrs

Glacial
gravels
220,000 yrs

London
clay
52 million yrs

98 m.y.

Chalk

Upper greensand and gault

144 m.y.

213 million years

Jurassic

Old red sandstone

Fig. 20.9. A geological map developed from the base map and extended downwards
for millions of years

Note: **This is a four-dimensional map showing length, breadth, depth and time**

Strata kindly dated by Dr Michael Frost

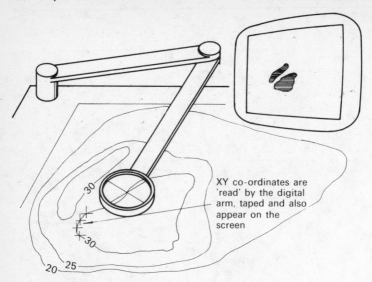

Fig. 20.10. The RD digital tracer

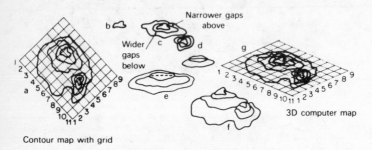

Contour map with grid

Fig. 20.11. To draw a three-dimensional contour map using a digital tracer

This map has equal x and y measurements. The computer can be instructed to draw it with increased or reduced vertical measurements to alter the effect.

The Axonometric, Hand-drawn Block Map

This 3D map is easy to make, especially if the reduced contour map with spacing lines is supplied by the teacher.

Method
1. Draw the contour map clearly to a **small** scale. This is easy with a reducing photocopier.
2. Tilt at 45°. Draw lines parallel to the bottom edges as shown and number the heights.

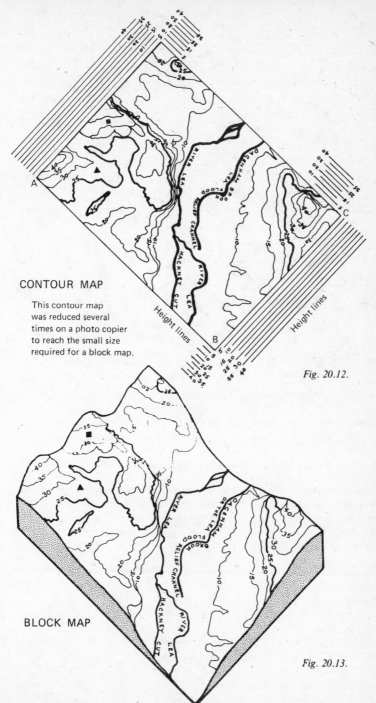

CONTOUR MAP

This contour map
was reduced several
times on a photo copier
to reach the small size
required for a block map.

Fig. 20.12.

BLOCK MAP

Fig. 20.13.

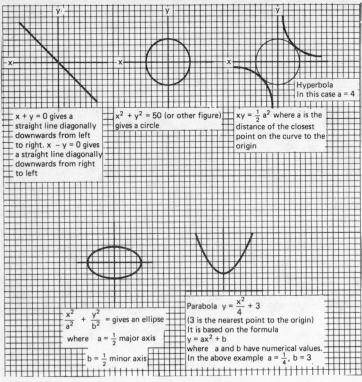

Fig. 20.14. Typical mathematical figures produced by formulae

3. Trace the bottom lines of the map AB, BC. Move them down to cover the 40 m lines.

4. Draw the 40 m contour.

5. Move the tracing lines up to the 35 m height lines and draw the 35 m contour. Do not go behind the 40 m contour.

6. Continue to move the tracing paper up and draw the contours in turn.

7. When all the contours have been drawn, join the ends of the contour lines smoothly to give the irregular outline. Complete the block and shade the vertical sides.

8. Number the contours to read from high to low.

You may like to revise the construction of Figs 20.12 and 20.13 and compare the different methods of drawing block maps. Block maps can be drawn in isometric projection, axonometric projection or even perspective, by hand and by computer.

Cartesian Curves

Fig. 20.14 shows typical Cartesian curves (after Descartes, the French philosopher and mathematician). They are found in all algebra textbooks but engineering draughtsmen hardly ever use them. Today, however, computers are becoming more important, and they draw by calculating Cartesian curves.

To plot Cartesian Curves Without a Computer

Each curve has its own formula. Each point on the curve, or straight line, is the crossing of an x and a y co-ordinate.

Method (which you may already know from studying algebra). Choose a series of values for x and calculate y. This will give a set of figures which are then plotted on the square grid paper.

Example

$$x + y = 0 \qquad \text{Let } x = -4 \ -3 \ -2 \ -1 \quad 0 \quad 1 \quad 2 \quad 3 \quad 4$$
$$\text{Then } y = +4 \ +3 \ +2 \ +1 \quad 0 \ -1 \ -2 \ -3 \ -4$$

When plotted, this gives the straight line in Fig. 20.14(*a*). Computers draw straight and curved lines in the same way, by calculating the co-ordinates of points on the path and plotting them in the correct positions.

The Light Pen

Glass fibres can be made to carry light. The light enters one end of the fibre and travels to the other. It tries to escape through the side walls but is reflected back into the fibre each time. A bundle of fibres can carry enough light to take photographs, for example, inside the human body.

A pen with a point of light at the end, can be used to instruct a television screen. The pen touches a point on the screen. This lights up and its position is also recorded in the computer memory. Other points are touched and the computer can be instructed to join them.

The pen can be used to choose a point on the screen. Then the computer can be instructed to draw a circle of a given size with that point as a centre, to draw a curve by calculation from that point or construct any figure. The light pen enables an engineer to sketch directly on the screen, alter his sketch, rotate it and turn it as he chooses and, when satisfied, have an exact drawing made by the computer.

An Electronic Pen

An electronic pen does the same type of work as a light pen. Writing or drawing with an electronic pen on the transmitter is automatically trans-

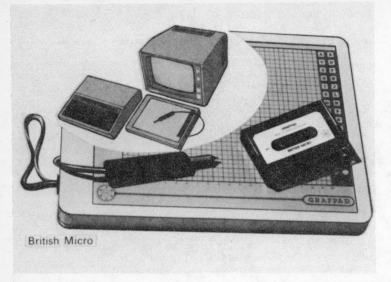

British Micro

Fig. 20.15. The Grafpad
By permission of Hegotron (Microcomputers) Ltd

mitted any distance to give an instantaneous copy on the receiver. Thus, for example, two engineers with transceivers could discuss details of an electronics circuit or other construction, drawing a diagram as they do so and amending it during the discussion, in such a way that each can see what the other has drawn and is left with a permanent copy at the end of the discussion.

Fig. 20.15 shows the complete system of pen, typewriter and VDU (Visual Display Unit).

The Grafpad

A graphic-tablet with a surface of 320 × 256 pixels (xy points) plus a menu area on the right. The positions of the pen on the tablet is transferred to the screen at up to 6000 xy points/second. Taped programmes allow circles, rectangles, etc. to be drawn; erasing; storage; or dumping to a printer.

Wire Frame Figures

Computer Aided Design programmes such as the 'Emsoft' used on the Sinclair Spectrum, can draw three-dimensional figures as 'wire frameworks'. The frameworks can then be rotated around different axes. An enhanced version of the programme can store a number of screen images and subsequently display these in rapid sequence to produce a moving display. Perspective is added by scaling the x and y co-ordinates by the value of the z co-ordinate.

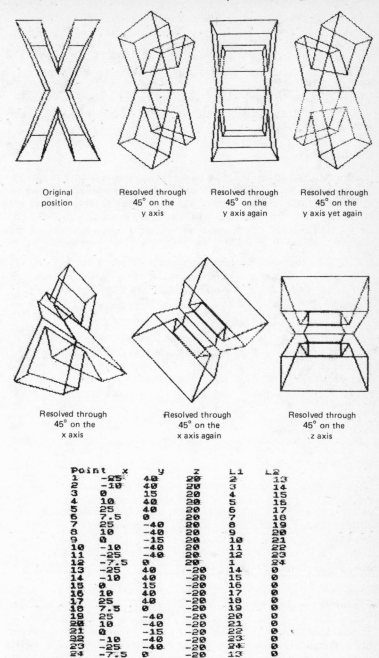

Original
position

Resolved through
45° on the
y axis

Resolved through
45° on the
y axis again

Resolved through
45° on the
y axis yet again

Resolved through
45° on the
x axis

Resolved through
45° on the
x axis again

Resolved through
45° on the
z axis

Point	x	y	z	L1	L2
1	-25	40	20	2	13
2	-10	40	20	3	14
3	0	15	20	4	15
4	10	40	20	5	16
5	25	40	20	6	17
6	7.5	0	20	7	18
7	25	-40	20	8	19
8	10	-40	20	9	20
9	0	-15	20	10	21
10	-10	-40	20	11	22
11	-25	-40	20	12	23
12	-7.5	0	20	1	24
13	-25	40	-20	14	0
14	-10	40	-20	15	0
15	0	15	-20	16	0
16	10	40	-20	17	0
17	25	40	-20	18	0
18	7.5	0	-20	19	0
19	25	-40	-20	20	0
20	10	-40	-20	21	0
21	0	-15	-20	22	0
22	-10	-40	-20	23	0
23	-25	-40	-20	24	0
24	-7.5	0	-20	13	0

Fig. 20.16. Wire frame figures
Printed by permission of 'Emsoft' of 37 Lennox Drive, Wakefield, West Yorkshire

The table in Fig. 20.16 shows the x, y and z co-ordinates of the 24 points in the original position. The L1 and L2 columns show where the lines go to from each position. Every time the wire framework figure moves, the computer recalculates the x, y and z position of each point. The L1 and L2 columns remain the same and always link the same points wherever they may be.

Only 24 vertices are shown but up to 500 can be specified on the 48K Spectrum. The programme may also be used in combination with a light pen or a digital tracer.

The NC-scriber

Rotring NC-scribers are designed to do the routine, repetitive parts of a drawing quickly and perfectly. They add lettering, dimensions and standard symbols to drawings. These shapes, which are stored in memory cassettes, can be repeated time and again, with accuracy controlled to 0.01 mm, without the machine tiring. The machine is a good work-horse which saves the time of skilled draughtspeople, releasing them for more creative work.

1 The NC-scriber showing the pen drawing a dimension (see Fig. 20.17).
2 The pen has drawn the symbol and returned to the starting position, ready to be moved to a new position and given the next order.

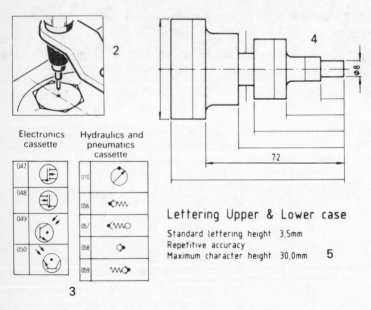

2

Electronics
cassette

Hydraulics and
pneumatics
cassette

047		010	
048		056	
049		057	
050		058	
		059	

3

4

$\phi 8$

72

Lettering Upper & Lower case

Standard lettering height   3,5mm
Repetitive accuracy
Maximum character height   30,0mm   5

Fig. 20.17. The NC-scriber
By permission of Rotring UK

3 A few typical symbols. The Wates' two bedroom bungalow (see Fig. 15.15) shows architectural symbols drawn with a scriber.
4 Dimensions and arrow-headed lines being added to a drawing.
5 Typical lettering.

Computer Mapping by Vertical Slices

Fig. 20.18 shows computer mapping by vertical slices, one of the latest methods of surveying. Each line shows the rise and fall in the landscape at one horizontal distance from the observer. These lines have been measured and mapped electronically. The picture can be thought of as a series of parallel cut-outs which make up the scenery in a toy theatre. Of course, any valleys behind hills are not seen. This sort of map can be taken from low-flying aircraft or, perhaps, ships.

Summary

Points and lines can be entered on television screens and into computer memories by typing in figures from a keyboard, with a digital tracer, with a light pen or with an electronic pen.

Lines and figures drawn as wire frameworks, can be rotated about various axes, enlarged or reduced, partial views taken, or parts altered. Thus car drawings, maps, etc. can be viewed from different directions and, when the best solution has been found, can be printed.

Mathematical curves can be calculated and drawn by computers. This does not mean that the computer will take the place of all geometrical

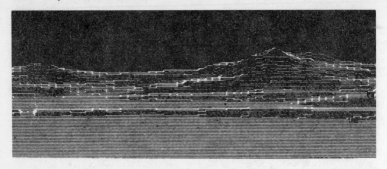

Fig. 20.18. Computer-drawn map
Published by permission of the Ministry of Defence and *The Times*

constructions. These are just as important as ever because, unless the geometry is understood, the computer cannot be used so effectively and certainly cannot be programmed.

Exercises

It is difficult to offer particular questions or exercises but a wide range of soft-ware is available on mini-computers and the choice is expanding rapidly.

21
Layout Techniques

Modern graphical designers use a great deal of ready prepared lettering. This is much quicker than hand lettering and anyone can do it. High quality hand lettering takes years to learn. A neat person can make good rub-down lettering at once. Because it is so easy, very badly spaced, unreadable and ugly lettering is often seen. Taste and sensitivity are vital.

Rub-down Lettering (Figs. 21.1–21.8 are printed by permission of Letraset Ltd.)

The lettering is printed on the underside of a sheet of transparent film. Thus it can be read, the correct way round, through the film. The letters on the underside are protected by a backing sheet. The process is very simple. Fig. 21.1 shows the letter 'b' being rubbed down in place with a spoon-faced tool. The letters of the word 'begin' have been rubbed down on to the paper from the film. In Fig. 21.2 they are now being burnished through a transparent sheet to hold them firmly in place.

Fig. 21.1. Rub-down lettering being applied

Fig. 21.2. Burnishing the lettering

Stick-down Architectural Tints

Figs 21.3 and 21.4 are like the rub-down letters but they consist of patterns, shapes, etc. printed on plastic which has a tacky back. The tints are

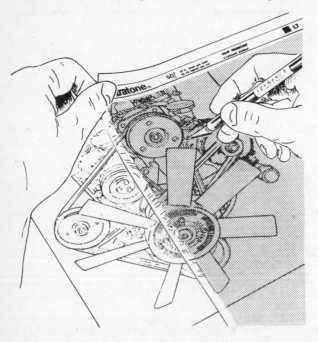

Fig. 21.3. Cutting a piece of plastic Letratone roughly to shape

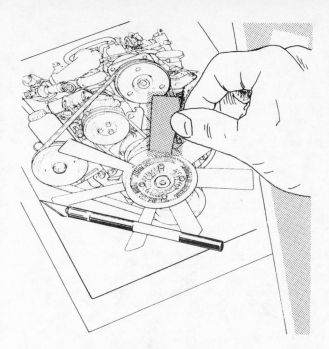

Fig. 21.4. Applying the architectural tint before cutting to exact shape

supplied with a backing sheet. The tint, still sticking to its backing sheet, is marked out and cut roughly to shape. The tint is peeled from the backing and stuck in place. It is then cut to the shape required and the waste peeled away. The sectioning on my drawings throughout this book has been done this way.

Fig. 21.5 shows a variety of tints being used over an ink drawing. In Fig.

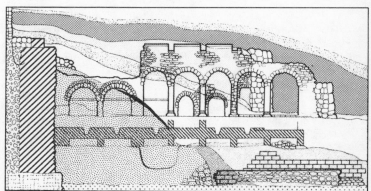

Fig. 21.5. A line drawing shaded with a variety of tints

Fig. 21.6. A tinted drawing using highlights

21.6 notice how the tints have been cut away to give the effect of sunlight on the castle walls, car bonnet, etc. This creative use of tints is found frequently in advertising drawing. Good examples are interesting to collect.

Fig. 21.7 shows an architect's perspective, softened with some rub-down trees (also Letratone). The glass has been rendered in a tint which fades from dark to light. One pane fades in both directions. This can be done by using a standard fade on one side and scraping off the other with a flat knife blade.

A beautiful cut-away drawing of a gear box is shown in Fig. 21.8, using a variety of tints and pen thicknesses. This drawing will teach you more than would a dozen descriptions.

Fig. 21.9 is an ink drawing enhanced with architectural tints, of a measuring table from the Roman city of Timgad, North Africa. Standard sized, hemispherical measuring bowls for olive oil and wine, have been cut in the stone, with drain plugs at the bottom. The largest bowl contained an amphora (26.26 litres) and the smallest, 1.54 litres. I made this drawing from a photograph some years ago. The isometric drawing of Brunel's bridge (see Fig. 7.41) and Choissy's dome (see Fig 7.40) also use tints.

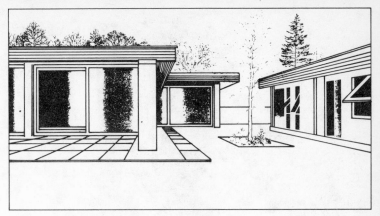

Fig. 21.7. A tinted architectural presentation drawing

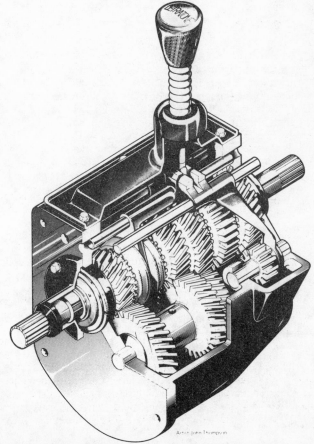

Fig. 21.8. Cut-away drawing of a gear box enhanced with tintwork
Artist: John Thompson

Fig. 21.9.

Colour

Modern examination papers often encourage candidates to colour drawings. This needs care and practice. Water-colour is not suitable for examination conditions although it can produce beautiful effects when time
is no object. Sheets of translucent colour tint similar to Letratone are used
by professionals, but these are very expensive and not suitable for examinations. **Wax crayons** are ideal for examination work. Avoid pastels as
they smudge. Choose colours carefully. Select four or five colours which
go well with each other and do not clash. If you are at all colour blind
(and seven per cent of boys and one girl in two hundred are colour blind
to some extent) ask someone to choose matching colours for you. Apply
the crayon evenly and lightly. Gentle tinting gives the best effect. Never be
strident.

Airbrush Work

An airbrush is a precision instrument which sprays fine droplets of paint
or ink on to a surface. The droplets dry almost instantaneously. The airbrush is connected by a hose to an air supply which can be a pump, an
aerosol, or a car tyre pumped up to 40 lb. per square inch pressure. Nozzles
can be adjusted to give larger or smaller paint droplets and to spray a fine
line or a broad band of colour. Cut card masks are used to give sharp
edges. The drawing is built up gradually and patiently, cleaning the reservoir for each colour change. The best airbrush work is printed on high-
gloss, smooth paper or Bristol board.

Figs 21.10 and 21.11 show two cut-away drawings which are easily
understood by non-engineers. The cut-aways reveal far more detail than
other drawings can show. These are beautiful examples of drawing and use

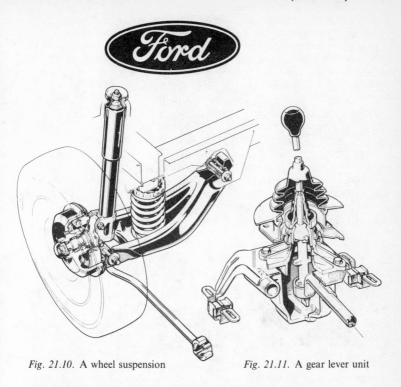

Fig. 21.10. A wheel suspension   *Fig. 21.11.* A gear lever unit

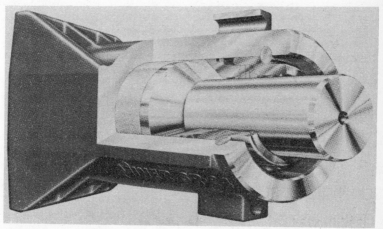

Fig. 21.12. A cut-away drawing of an engine bearing showing precise shading

a number of layout techniques. Some parts are shaded with architectural tints; some are pen drawn; and some are airbrush work. Examine the drawings carefully and decide which method has been used for each part.

An airbrush, itself drawn with an airbrush, is shown in Fig. 21.12. This is quite a large commercial instrument for heavy work. Fig. 21.13 shows a light airbrush being used to give delicate fading from one area to the next. Fig. 21.14 shows how precise shading can be. The whole drawing was shielded: tiny portions were exposed in turn and sprayed against a mask. The flat end of the round bar shows at least thirteen separate pieces of spraying, without counting the bevel or hole. Notice the method of showing a shiny surface by narrow strips of white. This is the standard method of illustrating highly polished metal. Airbrush work needs great skill and patience, as Fig. 21.14 shows. Notice also, the skilful drawing involved. The ellipses at the front are larger than those behind and at different angles, so that the whole drawing is in perspective.

Fig. 21.13. A Badger 250 airbrush
By permission of Badger Airbrush Co.

Badger Air-brush Replacement Parts

The drawings in Fig. 21.18 were made for identification by shopkeepers and users, not for manufacture. Notice the exploded layout of the parts showing the order in which they fit together. Each part has the code number of the airbrush (50 in this case), its own number and part name. Fig. 21.19 shows some enlarged drawings made from the Part Sheet.

Fig. 21.14 A Badger 400 airbrush
By permission of Badger Airbrush Co.

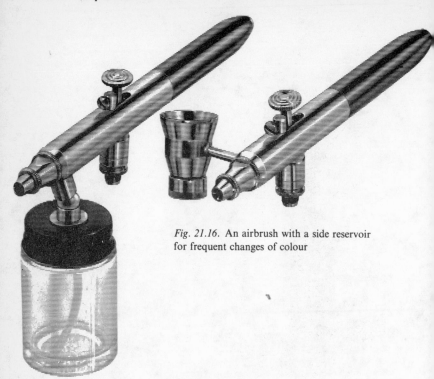

Fig. 21.16. An airbrush with a side reservoir for frequent changes of colour

Fig. 21.15. An airbrush with a glass paint reservoir below for covering fairly large surfaces

Questions

It is impossible to set questions on this chapter because most people will not have an airbrush, but it is a good opportunity to practise scale enlargements of simple drawings.

Draw as many of the parts shown in Fig. 21.18 as possible to twice or four times the given size (never three or five). Draw them one at a time. Put each one in a box with its airbrush number, part number, material and number off. Extra details of some of them may be found in the sectioned drawing Fig. 21.17.

Your drawings will not be nearly detailed enough for manufacture, as they will not show tolerances, machine details, surface finishing, plating, etc. or details of the materials to be used. They will be just exercises in simple scale enlargement and layout. Fig. 21.19 shows examples of the drawings which can be made from the amount of information.

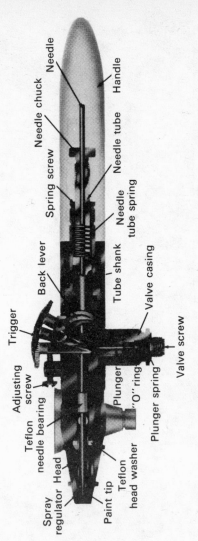

Fig. 21.17. Cross section of an airbrush

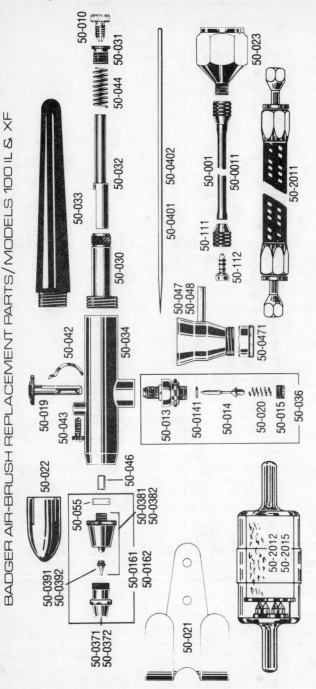

BADGER AIR-BRUSH REPLACEMENT PARTS/MODELS 100 IL & XF

50-010
50-031
50-044
50-032
50-033
50-030
50-042
50-034
50-019
50-043
50-022
50-046
50-055
50-0381
50-0382
50-0391
50-0392
50-0161
50-0162
50-0371
50-0372
50-021

50-0402
50-0401
50-001
50-0011
50-111
50-112
50-023
50-2011
50-047
50-048
50-0471
50-013
50-0141
50-014
50-020
50-015
50-036
50-2012
50-2015

Fig. 21.18. Drawings of airbrush parts

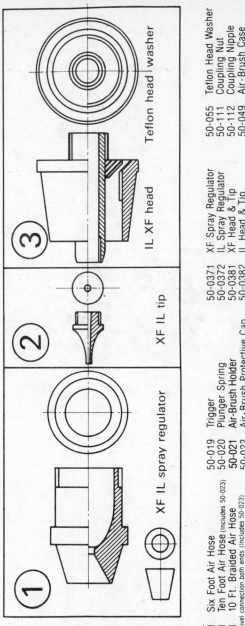

	XF IL spray regulator	XF IL tip	IL XF head	Teflon head washer

Part No.

50-001	Six Foot Air Hose
50-0011	Ten Foot Air Hose (includes 50-023)
50-2011	10 Ft. Braided Air Hose
	Swivel connection both ends (includes 50-023)
50-2012	Moisture Filter
	With 10 Ft. 50-2011 Hose
50-2015	Moisture Filter (only)
50-010	Needle Chuck
50-013	Valve Casing
50-014	Plunger & "O" Ring
50-0141	"O" Ring
50-015	Valve Screw
50-0161	XF Head Assembly
	Complete (50-0371-038l-0391-50-055)
50-0162	IL Head Assembly
	Complete (50-0372-0382-0392-50-055)

50-019	Trigger
50-020	Plunger Spring
50-021	Air-Brush Holder
50-022	Air-Brush Protective Cap
50-023	¼" Pipe Thread Fitting
50-0241	Jar Cover Gasket
	(3 Per Package)
50-025	Syphon Tube
50-030	Tube Shank
50-031	Spring Screw
50-032	Needle Tube
50-033	Handle
50-034	Shell (100) (w/Needle Bearing)
50-036	Valve Assembly
	Complete (50-013-014-0141-015-020)

50-0371	XF Spray Regulator
50-0372	IL Spray Regulator
50-0381	XF Head & Tip
50-0382	IL Head & Tip
50-0391	XF Tip
50-0392	IL Tip
50-0401	XF Needle
50-0402	IL Needle
50-042	Back Lever
50-043	Adjusting Screw
50-044	Needle Tube Spring
50-046	Teflon Needle Bearing
50-0471	Color Cup Cap
50-047	1/16 oz. Color Cup (100)
50-048	1/8 oz. Color Cup (100)

50-055	Teflon Head Washer
50-111	Coupling Nut
50-112	Coupling Nipple
50-049	Air-Brush Case
	(SPECIFY MODEL)
50-050	Prepared Beeswax

Fig. 21.19. Scale enlargements of some airbrush parts

22
Advanced Perspective Drawing

A Perspective Drawing Board

I once took out a provisional patent on a very simple perspective drawing board but Industry showed a complete lack of interest. Easy and cheap to make, it would be a valuable addition to a school drawing office and it is, of course, out of patent so anyone is welcome to make one.

Fig. 22.1 shows how a perspective drawing board is made. A piece of formica is glued in the centre of a piece of perforated hardboard. To use it, choose any pair of holes, one on each side and level with each other, as vanishing points. Pegs fit into these holes and a special rule pivots on them. The edge of the rule is in line with the centres of the rule holes. Once you have chosen the vanishing points, the rule works itself. Vertical lines are drawn with the T square. Measure all heights at the front corner. It can be fun to use.

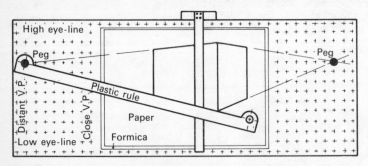

Fig. 22.1. A simple perspective drawing board

A Perspective Drawing Board using Arcs

A large board (see Fig. 22.2) has arcs of wood or cardboard screwed to it. The T square is made to run on the arcs, giving in effect, vanishing points well outside the board. The paper must be fastened on the board at a suitable level because the centre-line (eye-line) of the board is fixed.

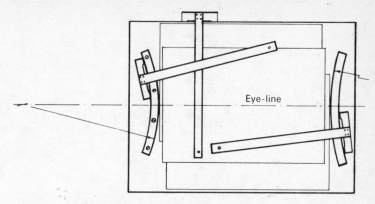

Fig. 22.2. A perspective drawing board using arcs

Three Point Perspective (see Fig. 22.3)

The same principles apply as for two point perspective. It is important not to put the vanishing points too close to the drawing or they give a distorted picture like a person's photograph taken from too close a position. The face becomes all nose. Fig. 22.4 shows that the vanishing points for Fig. 22.3 are more than the height of the drawing away

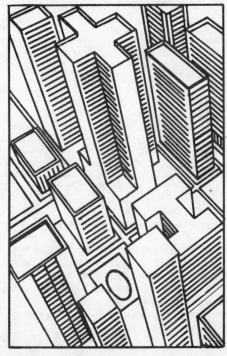

Fig. 22.3. Three point perspective

(*Fig. 22.4* (over) shows suitable positioning.)

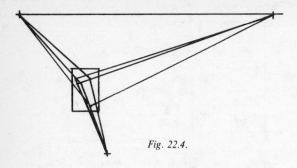

Fig. 22.4.

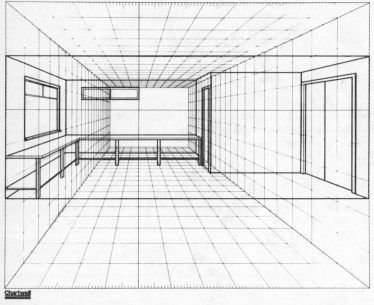

Fig. 22.5. Single point perspective grid with middle viewpoint
By permission of H. W. Peel & Company Limited

Perspective Grids

Figs 22.5 and 22.6 show the use of pre-printed perspective grids. Chartwell, part of H. W. Peel and Company Limited, have kindly printed these for me in black but normally they print in 'drop out' or 'transparent blue' lines. A drawing is made in black on the blue grid, but when the sheet is photographed, only the black drawing appears because the film cannot 'see' the blue. The blue is photographically transparent and 'drops out'.

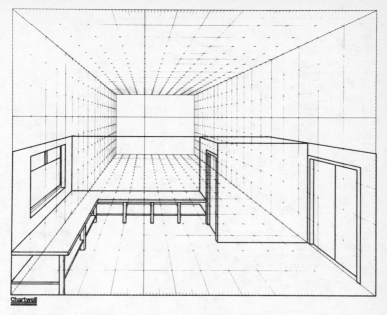

Fig. 22.6. Single point perspective grid with high viewpoint
By permission of H. W. Peel & Company Limited

Single Point Perspective Grid

The ready-drawn grid has perspective 'squares' on the ceiling, floor and two sides, with a vanishing point at the end. The only true sizes are at the front edges (see Figs 22.5 and 22.6).

Choose a viewpoint (the middle in Fig. 22.5 and a high view point in Fig. 22.6). Any other level could be chosen. The vanishing point remains the same but the amount of floor space alters. The grid is very flexible in use.

Two Point Perspective Grid

The grid represents three planes passing through each other, a horizontal plane, a front vertical plane and an end vertical plane. These are shown in Fig. 22.7. One would not be able to draw on the pointed corners, so only the parts inside the thick line in Fig. 22.8 are printed on the grids. Secondly, only two quarters of each plane is printed on the grids. In the front vertical plane in Fig. 22.7, there are four parts, A, B, C and D. Part A has been left out in Fig. 22.8 because it would confuse the two planes behind.

To use the grid The problem is to find where to start. It is a good idea to pin the grid to a board with a piece of tracing paper over it. You can try out the position without wasting grid paper. When you have the position right, work on the grid.

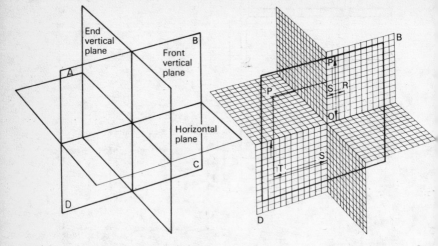

Fig. 22.7. Three intersecting planes   *Fig. 22.8.* A perspective grid

It is easy to find points on quarters B and D by measuring up and across. To find point P on the missing part of the grid, Measure up the height on grid B and project the grid line forward. Measure the distance forward at ST and project up to the cut at P.

Fig. 22.9 shows a modern three-storey town house with garage in Third Angle projection. It is drawn on a square grid so that one 'square' equals one foot. Some of the perspective 'squares' are large and some are small but each one represents a foot.

Fig. 22.10 shows the house drawn in perspective on a grid.

Method Cover the grid with tracing paper for the first try. **Start with the nearest corner ab** (see Fig. 22.11). Choose a suitable position. If it is too high, you will not see the roof terrace. If too low, the front wall will be over shortened. Counting the squares can be difficult. Line ab is 21 squares high. Count from a up to 12 on the 'cliff edge' where the planes meet. Creep carefully along the cliff edge to 0 and climb again to d (12 + 9 = 21). Project line cd to b. Similarly, measure the width as ae, up to d and across to c.

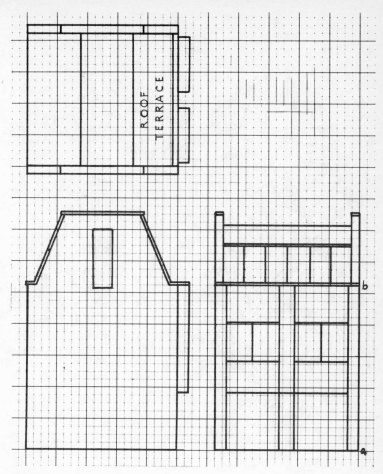

ROOF TERRACE

Fig. 22.9.

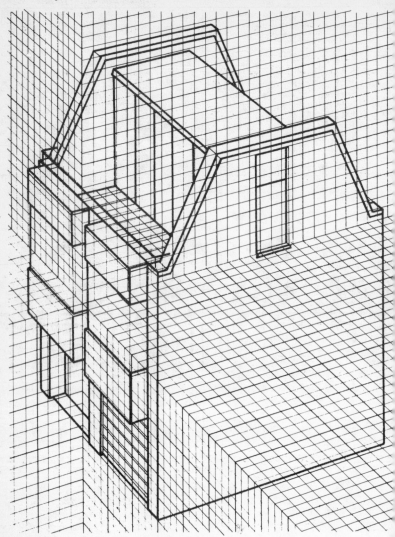

Fig. 22.10.

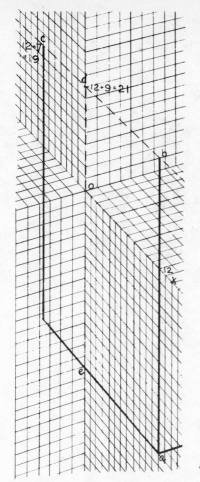

Fig. 22.11.

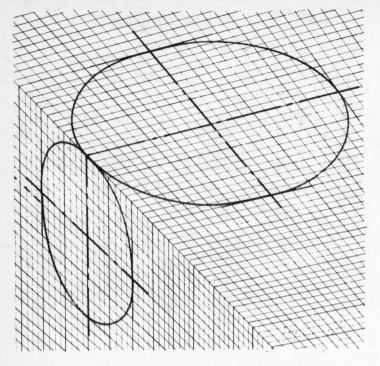

Fig. 22.12.

Circles Fit into 'Perspective Squares'

As the ellipses get further away, the 'squares' become smaller and so do the ellipses (see Fig. 22.12).

There are three point perspective grids but they are seldom used except in aeroplane and science-fiction illustrations.

To make a Perspective Grid for a Near Object

In Fig. 22.13 part (*a*) is a rough sketch of a block-shaped building which is to be drawn in perspective. The vanishing points will be far outside the edges of the paper so we must construct a grid. In part (*b*), it is easy to draw the nearest vertical corner ab and three of the sloping edges ad, ae and bg. The fourth line is more difficult. By guesswork, it could be bh[1], bh[2] or bh[3]. The solution, in part (*c*), is to draw the sloping lines ad, ae and bg. Then draw a vertical rectangular plane parallel to the viewer. Take any point 0 in ae. Drop a vertical OP to bg. Draw OR horizontally to reach ad. Complete rectangle ROPQ. The fourth sloping perspective line must be bQ extended. The grid can be completed as explained in Fig. 10.30 by repeatedly finding the perspective centres of the rectangles, drawing ver-

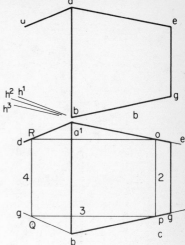

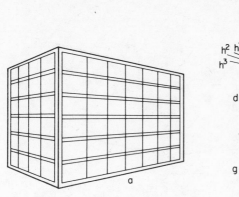

Fig. 22.13. Constructing a perspective grid

ticals through them and radiating lines which converge towards the vanishing points.

Perspective Shadows

Shadows are quite easy to draw and help to make drawings more realistic. Some artists completely ignore shadows. Lowry said that they only confuse things, but others have used them dramatically. They are simple if you consider them in stages.

Fig. 22.14 shows a series of posts and light coming from a street lantern. The posts cast thin triangles of shadow. We see only the line of shadow on the ground, but if you passed your hand through the triangle, a line of shadow would cut it. In Fig. 22.15 a flat roof on the posts now casts a shadow abcd. In Fig. 22.16 the pitched roof takes the shadow out to g and the inside of the hut becomes dark.

The plan and elevation of a pillar and a cylinder cast shadows (see Fig. 22.17). This time the light comes from the sun, so all the rays are parallel. The cylinder casts a circular shadow as if it was in oblique projection. Architects often make their presentation drawings like this. They are drawn to show a proposed building in a street scene, with a few elegantly dressed passers-by and a Rolls-Royce. Carefully coloured, with deep shadows, they are advertising material, not fact, and should be regarded with suspicion. They can, however, be very attractive.

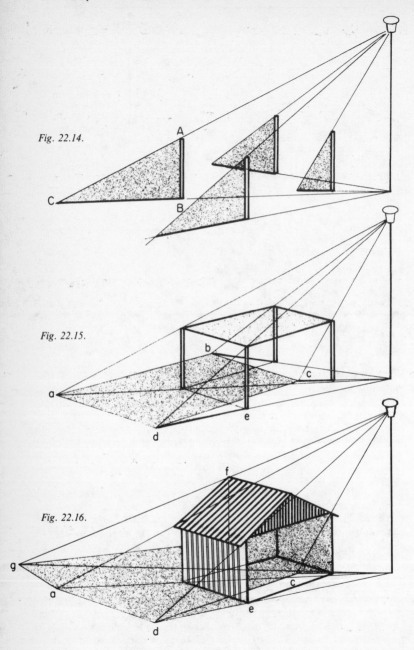

Fig. 22.14.

Fig. 22.15.

Fig. 22.16.

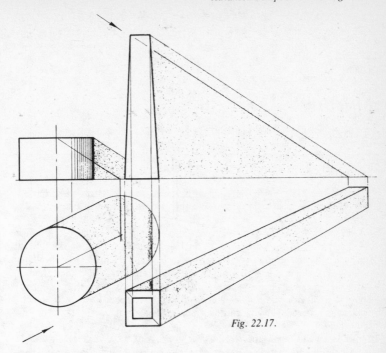

Fig. 22.17.

Questions

Copy Figs 22.14–22.17 and then make up some drawings showing perspective shadows of your own. Buy some perspective grids and experiment with them. A set of mixed grids gives a very interesting variety of drawings. Alternatively, make the grids yourself. The great secret is to go on experimenting. Add shadows to your earlier drawings and make new ones with shadows.

Appendix 1
Rules for Examinations

1 Make sure that you are familiar with your drawing instruments. Do not buy a new compass the night before the examination, or it may slow you down.
2 Always try to tackle the full number of questions on the examination paper. Normally, all questions carry the same number of marks. Divide the time by the number of questions to be answered, leaving about ten minutes at the end for revision. Try not to exceed this allocated time on any one question. If you spend all the time on two questions out of say five, you cannot possibly gain enough marks to pass. Even setting out a question correctly, will gain some marks, so at least start the full number.
3 Make sure you have answered as many parts of the question as possible. If a machine drawing asks for: scale, projection, and six important measurements; make sure that you put them in. They carry a set number of marks. Put them in even if you have not completed all the views. They are the quickest earned marks on the paper.
4 Draw neatly and **leave in all construction lines** unless you are told to rub them out.
5 Line in clearly and neatly. **Make sure the examiner can see the lines easily**.
6 Make sure you come to the examination room at the right time on the correct day and with all your equipment. Nothing can be done if you are missing.

Appendix 2
Examination-type Questions

The pages 377–87 show examples of the sorts of questions asked by different examiners. Some examination boards supply plain drawing paper for the examination. Others pre-print some parts of the answers so that the outlines of views, or starting points for views, are given. Examples of both sorts are given. Make sure that you are familiar with the style of examination paper and question usually set **by your particular board**. The style can change suddenly when a new chief examiner is appointed, but recent papers are the best possible guide.

Questions 1–4 (see Figs A1–A4)

These are typical examples of pre-printed examination papers. The candidate fastens the paper on the drawing board and completes the questions. In this way, the examiner can test the candidates' knowledge quickly: a dozen or fifteen questions may be asked in thirty minutes. Not all questions carry the same mark.

Questions 5–8 (see Figs A5–A8)

Question 9

In Fig. A.9 the plan and elevation of a cone is given. Draw these views. Then draw the plan, elevation and end elevation, when the base has been lowered to lie in the horizontal plane. (You may be surprised by the result – see Fig. A.17.)

Finding Areas by Integration

To integrate is to combine parts into a whole. It is often difficult to calculate the area of an irregular shape. Drawing a solution is often easier. The method is to cut up the shape into parallel strips and join them into one block (to integrate them).

Question 10

Fig. A.10 shows an estuary with an island. Find the area of water by integration. (See Fig. A.19 for solution.)

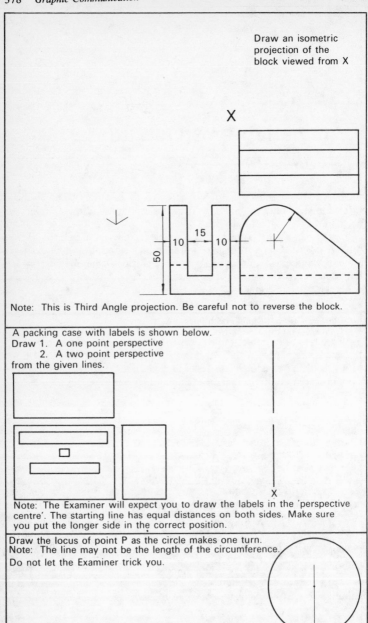

Draw an isometric
projection of the
block viewed from X

X

Note: This is Third Angle projection. Be careful not to reverse the block.

A packing case with labels is shown below.
Draw 1. A one point perspective
     2. A two point perspective
from the given lines.

X

Note: The Examiner will expect you to draw the labels in the 'perspective
centre'. The starting line has equal distances on both sides. Make sure
you put the longer side in the correct position.

Draw the locus of point P as the circle makes one turn.
Note: The line may not be the length of the circumference.
Do not let the Examiner trick you.

P

Fig. A.1. Pre-printed examination-type paper – sheet 1

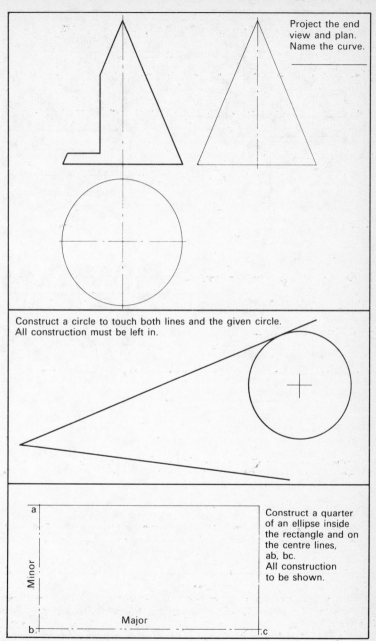

Project the end
view and plan.
Name the curve.
_____

Construct a circle to touch both lines and the given circle.
All construction must be left in.

a

Minor

b   Major   c

Construct a quarter
of an ellipse inside
the rectangle and on
the centre lines,
ab, bc.
All construction
to be shown.

Fig. A.2. Pre-printed examination-type paper – sheet 2

380   *Graphic Communication*

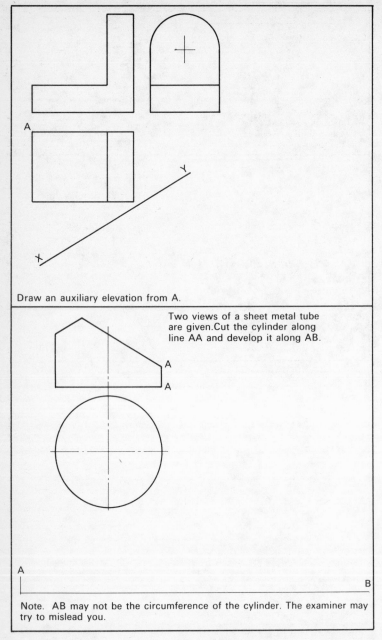

Draw an auxiliary elevation from A.

Two views of a sheet metal tube are given. Cut the cylinder along line AA and develop it along AB.

Note. AB may not be the circumference of the cylinder. The examiner may try to mislead you.

Fig. A.3. Pre-printed examination-type paper – sheet 3

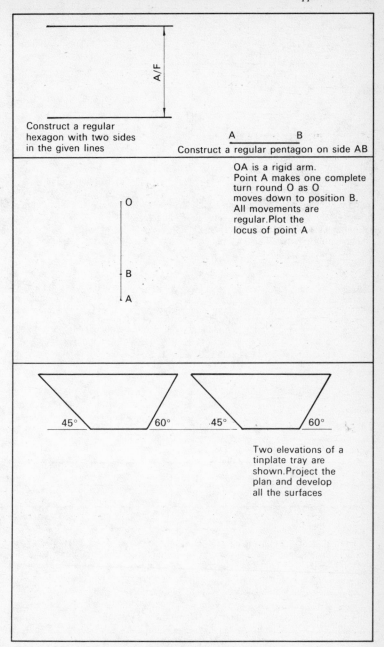

A/F

Construct a regular
hexagon with two sides
in the given lines

A   B

Construct a regular pentagon on side AB

OA is a rigid arm.
Point A makes one complete
turn round O as O
moves down to position B.
All movements are
regular. Plot the
locus of point A

O

B

A

45°   60°   45°   60°

Two elevations of a
tinplate tray are
shown. Project the
plan and develop
all the surfaces

Fig. A.4. Pre-printed examination-type paper – sheet 4

382   *Graphic Communication*

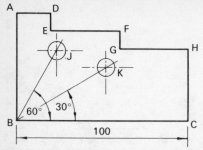

Fig. A.5. Testing of measurement skills. Time 10 mins.

Question:
AB = 60, BC = 100, DE = FG = 10, BJ = 50, BK = 90
Set out and add **datum** measurements
Measure from AB and BC
Fig. A.6. Another quick test. Time 4 mins.

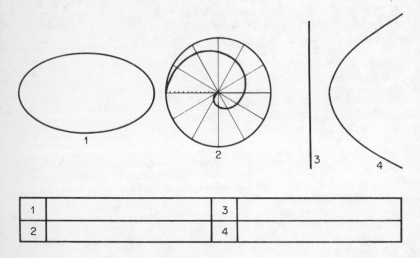

1		3	
2		4	

Question:
Name the curves. PRINT answers in the box.

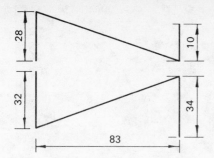

Fig. A.7. True length question. The plan and elevation of a wire are shown. There are three true lengths here. Time 4 mins.

Fig. A.8.
Question:
The plan and elevation of a bent wire are shown.
Find the true length of the bent wire by projection and stepping off

This is a test of accuracy.

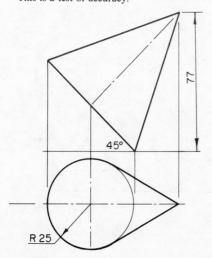

Fig. A.9.

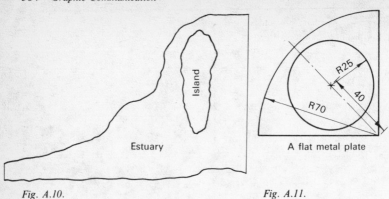

Estuary

Island

A flat metal plate

Fig. A.10.   *Fig. A.11.*

Question 11

Fig. A.11 shows a flat quadrant of metal with a lightening hole in it. Find the area by integration.

The area of Fig. A.10 is difficult to calculate but the area of Fig. A.11 is easy. How would changing the position of the circle in the quadrant affect the area in Fig. A.11? (See page 390 for solution.)

Question 12 'Match the Drawings'

Fig. A.12 is a quick method of testing if the candidate understands translating picture drawings into orthographic views.

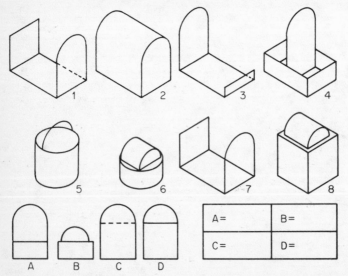

Fig. A.12. A 'match the drawings' question. These are very common. Ask your friends to test you.

Fig. A.13. Auxiliary projection – question 1.
(For solution see Fig. A.19.)

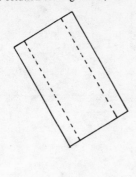

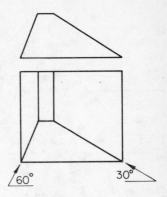

Fig. A.14. Auxiliary projection – question 2.
(For solution see Fig. A.20.)

Questions 13 and 14 on Auxiliary Projection

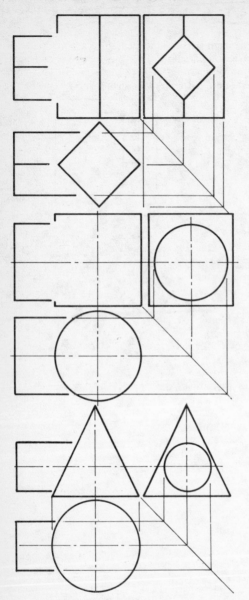

Fig. A.15. Pre-printed intersection questions. Complete the lines of intersection.

Question 15 on Intersections

Examiners often set out the main parts of the views on the actual answer paper and ask for the intersection lines only. This allows them to ask more questions in the time. The three questions in Fig. A.15 could be answered in the time taken to set out and a single drawing in the normal way.

Question 16

Some questions depend on careful reading and re-reading of the question. This is an example: there is no drawing so you should make thumbnail sketches as you are thinking out the solution.

A tetrahedron has a vertex D and stands on its base ABC which is in the horizontal plane. The edge AC (which is parallel to the vertical plane in the plan view) is 120 mm long. BC equals 100 mm and angle ABC is 90°. The vertex D is vertically above the centre of the circle which could be inscribed in the base, and DC is 105 mm long.

In First Angle projection draw:

(*a*) The plan and elevation of the tetrahedron.
(*b*) A second elevation as seen from the left of (*a*).

Determine (i) the true length of AD, (ii) the true inclination of face ABD to the horizontal.

See Fig. A.21 for solution.

Question 17 A Strange Ellipse Problem

A final question which you are most unlikely to be asked in the examination, but which you may like to puzzle over. It will defeat your friends at first. Draw any ellipse. Step off points 5 mm apart all round the perimeter. Draw a circle from every point to pass through f^1. What do you find and why?

See Fig. A.22 for solution.

Solutions to selected questions from Appendix 2 are given on pages 388–95.

Appendix 3
Solutions

Question 9 (see Fig. A.17)

Question 10

Method of integration. (See Fig. A.18.) Draw any convenient base line near one edge and divide it into vertical strips 0, 1, 2, etc. (see Fig. A.18(a)). Draw the mid-lines $\frac{1}{2}$, $1\frac{1}{2}$, $2\frac{1}{2}$, etc. Rectangle abcd, is drawn through the top and bottom of the middle line $1\frac{1}{2}$. It is about equal in area to the strip of water 1–2.

Draw a new base line and extend it to a convenient point P (see Fig. A.18(b)). Project down from the **mid-lines** in Fig. A.18(a). Measure the lengths of the mid-lines in Fig. A.18(a) **over the water**. It needs to be known how much water is covered by each mid-line.

Transfer these distances to drawing Fig. A.18(b). Measure each one up from the base on its mid-line as at $\frac{1}{2}$ and $6\frac{1}{2}$. Project horizontally from each of these points to the vertical OQ. Join each to P. These slopes represent the rises of the mid-points.

Draw from 0 to 1 parallel to P$\frac{1}{2}$ in B (see Fig. A.18(c)). Repeat for 1–2 parallel to P1$\frac{1}{2}$, etc. Notice that the length $6\frac{1}{2}$ in Fig. A.18(b) is less than $5\frac{1}{2}$, so the slope from 5 to 6 in Fig. A.18(c) is less steep than the ones on either side.

The area of water = OP X the final height 7H. If you choose a longer or shorter OP line, the final height 7H will be shorter or longer, but the final area will stay the same.

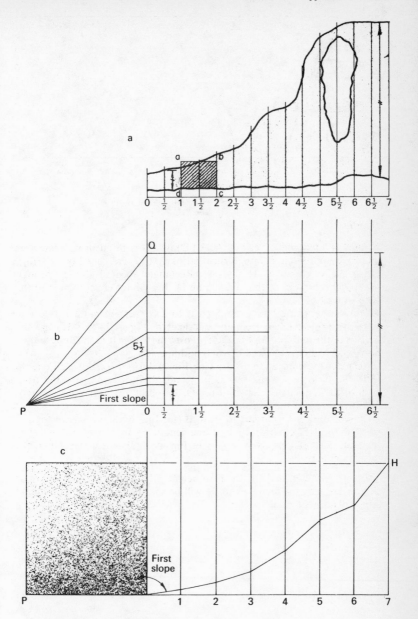

Fig. A.18. Solution to Question 10

These three views would be drawn one on top of the other in fact, but have been separated to make the construction clear.

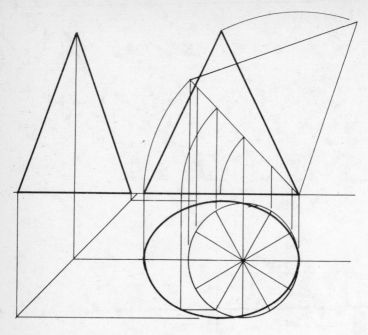

Fig. A.17. Solution to Question 9

Question 11

The mathematical solution is:

$$\text{Area} = \frac{\pi \times 70}{4} - \pi \times 25$$

It could also be solved by integration. Moving the position of the circle in the quadrant would not affect the result.

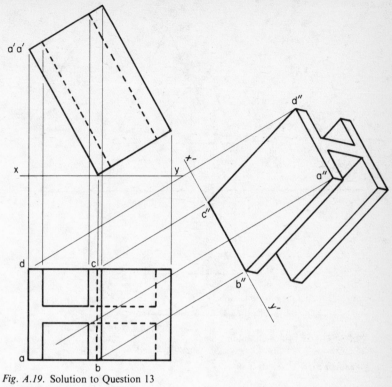

Fig. A.19. Solution to Question 13

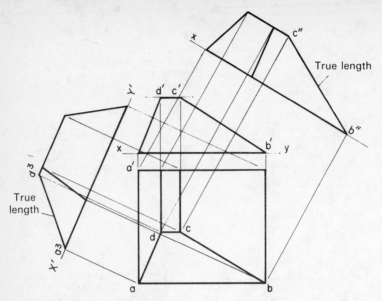

Fig. A.20. Solution to Question 14

Question 14 (see Fig. A.20)

Question 16

Solution Follow in the preliminary sketches (see Fig. A.21(*a*)) and the final drawing (see Fig. A.21(*b*)). Base ABC lies horizontal. AC is parallel to the vertical plane. Draw AC. Angle ABC = 90° so it must fit in a semicircle. Draw a semicircle on AC. Mark off BC = 100. Then ABC is the base. Inscribe a circle in the base. Project centre D to the front elevation. We do not know the vertical height but we know the true length of CD. In plan, swing CD about C to bring it into AC. Project up. In elevation, strike off C′D′ = 105 mm. Draw horizontally from D′ to D″. Complete the elevation. Project the inclination of ABD to the base as shown.

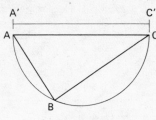

Setting out the base

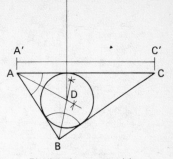

Finding the apex position

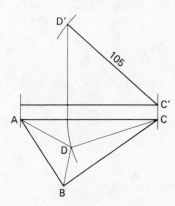

Swing CD parallel to the
XY line and project up.
Strike off the true length
of CD (105 mm)

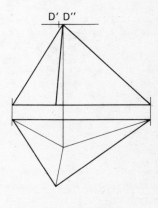

Project horizontally from D'.
Project up from D to D''.
Line in.

Fig. A.21. Solution to Question 16
(*a*) Preliminary sketches

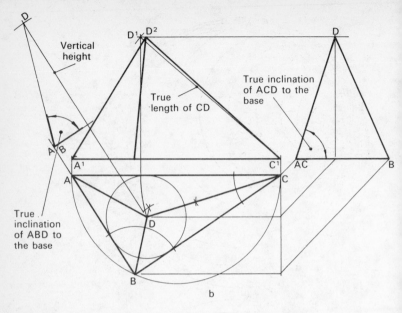

Vertical
height

True
length of CD

True inclination
of ACD to the
base

True
inclination
of ABD to
the base

b

(*b*) Finished projection

Question 17

The solution is shown in Fig. A.22. Marios Andreou was a student of
mine who stumbled on this, so we called it the 'Marios Andreou Constant'.

Given an ellipse with major axis AB, minor axis CD and focal points f', f^2
as shown in Fig. A.22. **Then** any circle with its centre on a point on the
ellipse and radius from this point to f^1, will be tangential to the circle
centre f^2 and radius equals the major axis.

Proof Draw the ellipse and circle centre f^2 radius equal to the major axis
AB. Draw circle centre A and radius Af^1. $Af^1 = Bf^2$ so circle centre A will
touch circle centre f^2. Circle centre B is similar. Let E be any point on the
ellipse. $Ef^1 + Ef^2 =$ the major axis $= Ef^2 + EQ$.

What would happen with a parabola?

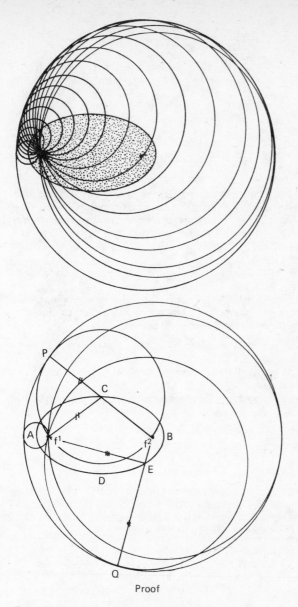

Fig. A.22. Solution to Question 17. The 'Marios Andreou Constant'.

Glossary

Accuracy This is the secret of man's control over the world. The examiner will expect your drawings to be accurate to less than a millimetre, but this is very crude. Today, scales are made accurate to seven microns (seven millionths of a metre) over a length of half a metre and at a standard temperature.

Architectural tint. (See index.)

Auxiliary view A helping view – one which makes an object easier to understand.

Axonometric (or Planometric) A form of picture drawing where the plan is drawn true shape but may be turned in any direction. Vertical lines stay vertical. (Axon–axis – drawn round a vertical axis.)

Cabinet projection A form of oblique projection in which thickness lengths are halved.

Cavalier projection A form of oblique projection where thickness lengths are drawn full length.

Concentric circles Circles having the same centre. (See Eccentric circles and Fig. G.1.)

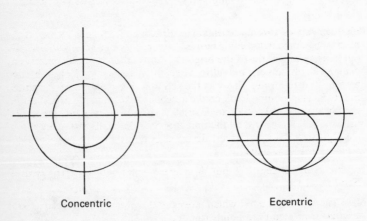

Concentric   Eccentric

Fig. G.1.

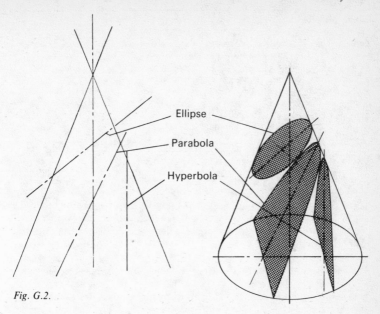

Ellipse

Parabola

Hyperbola

Fig. G.2.

Conic sections Cones cut by planes give circles, ellipses, parabolas or hyperbolae. (See Fig. G.2.)

Cycloid The locus of a point on the circumference of a circle as the circle rolls one complete turn along a straight line. (For varieties of cycloids see Index.)

Diagonal A straight line joining two corners (not adjacent) of a plane or solid figure.

Digital copying of rare documents The British Library first identified the need for equipment that could be used to reproduce contents of books without causing any harm to the originals. Conventional photocopiers may damage book spines, and ultra violet light causes ancient print and writing to deteriorate. This was the solution. See diagram of a scanner, Fig. G.3. The rotating book cradle accepts books or single pages. Books need be opened only to an angle of 80°, which protects their spines, and they are fully supported all the time. Fig. G.3 illustrates the operation of the Optronics Digital Copier, designed by Optronics Ltd., Cambridge Science Park, England.

Eccentric circles Eccentric – out of centre. Circles with different centres. (See Fig. G.1.)

Ellipse The locus of a point which moves so that it is always the same **total** distance from two fixed points (the foci).

Focus – plural foci. A fixed point round which other points move.

Greek alphabet.

Letters		Name	Letters		Name
A	α	alpha	N	ν	nu
B	β	beta	Ξ	ξ	xi
Γ	γ	gamma	O	o	omicron
Δ	δ	delta	Π	π	pi
E	ε	epsilon	P	ρ	rho
Z	ζ	zeta	Σ	σ	sigma
H	η	eta	T	τ	tau
Θ	θ	theta	Y	ν	upsilon
I	τ	iota	Φ	φ	phi
K	κ	kappa	X	χ	chi
Λ	λ	lambda	Ψ	ψ	psi
M	μ	mu	Ω	ω	omega

Holography A system of photography which gives a three-dimensional effect on a flat film. A beam of lazer light (which has only one wave length) is split into two beams. These strike the object from different angles and are both reflected on to the same photographic film. The differences between the pictures form fringe patterns which the viewer sees as a three-dimensional picture. As the viewer moves, he can see the object from different angles. A master hologram is very expensive but it can be printed on foil when it becomes an embossed hologram. These can be cheap and in a few years could become very cheap indeed.

Hopper Box with tapered sides.

Hyperbola A double curve produced by a point moving so that it is always further from a focal point than from a straight line (directrix) in a fixed ratio.

Hypotenuse The longest side of a triangle.

Inclined plane A plane inclined to *either* the VP or the HP.

Involute The path of a point moving away from a point which moves round and away from a point or figure in some regular motion.

Isometric projection. (Iso – equal, metric – measurement.) A form of picture drawing where vertical lines stay vertical and horizontal lines slope upwards at 30°. All lengths are drawn to true length unless an isometric scale is required. This is rare at GCSE level.

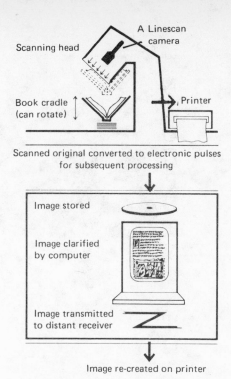

Scanned original converted to electronic pulses
for subsequent processing

Image stored

Image clarified
by computer

Image transmitted
to distant receiver

Image re-created on printer

Fig. G.3. Copying of rare documents

Landscape A picture which is wider than it is tall.
Locus plural loci – The path of a point.
Logo A symbol standing for an organisation or an idea.
Logotype. (Logus–word.) A single piece of printers' type bearing a non-heraldic design chosen as a badge of an organisation and used in advertisements or on notepaper, etc.

Non-carbon copying papers Each sheet is backed with a layer of microcapsules which contain a dye. Under the pressure of a pen or typewriter key they burst on to the special china clay coating on top of the next sheet and a chemical reaction produces a copy image.

Oblique plane A plane inclined to *both* the VP and the HP.
Oblique projection A form of picture drawing where the front elevation of **the front surface** is projected back at 45°.
Orthographic projection A system of drawing where one view is projected (thrown) from one plane to others at right angles and the planes are then folded flat. Also known as right-angled drawing.

Parabola The locus of a point which moves so that it is always an equal distance from a focal point and a straight line (directrix).

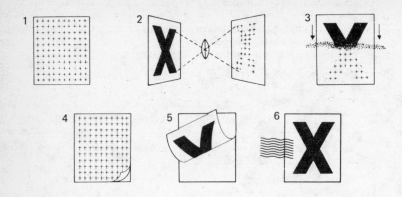

Fig. G.4. How xerography (photocopying) works

Perspective A system of drawing where sets of parallel lines are drawn to meet at vanishing points.

Photocopying (copying by light). In 1938, Chester Carlson invented xerography (dry writing) using two well-known facts:

1 Materials with opposite electrical charges are attracted.
2 Certain materials become better electrical conductors when exposed to light.

These phenomena allowed him to make good, fast, cheap copies on plain paper. Before this, most copying had been a wet process and had required special, sensitised papers. Fig. G.4 shows how xerography works.

Basic Xerography (one form of photocopying)

1 A photoconductive surface is given a positive charge (+).
2 The image of the document is exposed on the surface. This causes the charge to drain away from the surface in all but the image area, which remains unexposed and charged.
3 Negatively charged powder is cascaded over the surface. It electrostatically adheres to the positively charged image area making a visible image.

4 A piece of plain paper is placed over the surface and given a positive charge.
5 The negatively charged powder image on the surface is electrostatically attracted to the positively charged paper.
6 The powder image is fused to the paper by heat.

After the photoconductive surface is cleaned, the process can be repeated.

Some photocopiers can reduce and enlarge in fixed proportions to suit the A paper range. A4 can be enlarged to A3 or reduced to A5. Some photocopiers have zoom lenses which allow any change of size. These are particularly useful when enlarging and reducing, for example, a set of different maps of the same district to the same scale.

Enlargement and Reduction by Photocopier

Most photocopiers reduce and enlarge in standard proportions. For example, A4 is half the **area** of A3. Therefore the length of side is reduced by about 71%.

$$\frac{71}{100} \times \frac{71}{100} = \frac{5041}{10,000} = \frac{1}{2}$$

The standard reductions and enlargements vary from 65% to 154% **length of side**. Some photocopiers have zoom lenses and are infinitely variable between these limits.

Photography. Photography is used in the drawing office to:

1 Store drawings on tiny microfiche which save the huge plan chests needed for full sized drawings
2 Restore old and battered drawings – the creasings and yellowing disappear.
3 Update drawings by printing them on wash-off film. Then, with a soft eraser and water, the unwanted lines and figures are removed and new ones can be drawn in. The majority of the drawing can be re-used – a great saving of time. (See Fig. G.5.)
4 To photograph printed circuit diagrams, correct them, update them, etc. Then a reduced size film negative can be used to etch away the unwanted metal foil and so make the actual printed circuit. The drawing becomes the photograph, which in turn becomes the etching tool. (See Fig. G.6.)

Plane A flat surface with length and breadth but no thickness.
Planometric drawing alternative name for axonometric drawing.
Portrait A picture which is taller than it is wide.
Produced (in geometry). Extended – made longer.

WHAT IS THE DIFFERENCE?

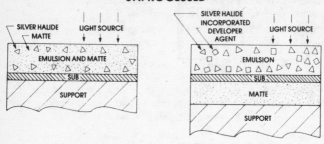

CONVENTIONAL FILM   WASH-OFF FILM

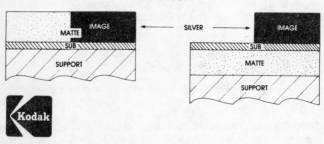

Fig. G.5. Matte is the drafting surface on ESTAR base (polyester) 0.1 mm thick

Shadows Architectural draughtsmen often add shadows to make their drawings appear 'real' to the client. Some drawings give very false impressions. The original drawings of the House of Commons make the window buttresses appear far deeper than they are. The builder could have been accused of not supplying hundreds of tons of stone. Do not be misled.

Sine – trigonometry – The ratio of the perpendicular subtending an angle to the hypotenuse. (Abbreviation sin – see also cosine.)

Subtend (of a chord or side of a triangle). To be opposite to an arc or angle.

Tangent. (Abbreviation tan.) A line which touches but does not cut another line even if produced (geometry). The ratio of the perpendicular subtending an angle in any right angled triangle to the base (trigonometry).

Tolerances Exact measurements are often given on drawings but it would be much too expensive to insist on perfect sizes. The cost of manufacture and measurement would be enormous. Instead, 'tolerances' are set. A part is accepted if it is within a narrow band of sizes, perhaps 0.002″ or 0.05 mm larger or smaller. Accuracy is improving all the time. James

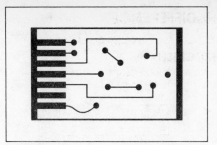

Start with your master etched circuit layout.

Step 1: Make a positive reproduction on wash-off film.

Step 2: Wet-erase unwanted details and tap in new leads.

Step 3: Photograph revised drawing on film.

Step 4: Make a reduced-size film negative (used to expose sensitized metal laminate).

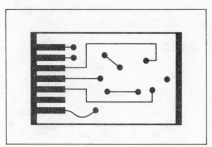

Step 1

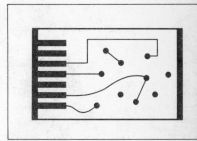

Step 2

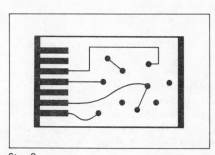

Step 3

Step 4

Fig. G.6.

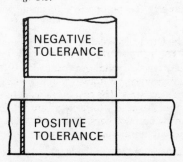

NEGATIVE TOLERANCE

POSITIVE TOLERANCE

$\varnothing 19.98 - 0.02$

$20.00 + 0.02$

Watt was pleased when he received a cylinder accurate 'within the thickness of a thin shilling'. Today, the Japanese machine their railway carriage axles to within 3 microns.

Fits and Tolerances are listed in detail by British Standards. The ISO system is accepted by Britain, much of Europe, Canada and USA. Basic Principles: two mating components have a **basic size**; the tolerances are graded – the smaller the tolerance, the lower the grade.

Fits are listed for different purposes, varying from very slack running to a heavy press fit where parts are to be forced together and never separated.

Shafts have a negative tolerance and holes have a positive tolerance. This ensures that the shaft will enter the hole. (See Fig. G.6.)

Vertex The topmost point.

Wash-off film. (See Photography.)

xy line The line where two planes intersect.

Short Bibliography

The books listed here are not in any way compulsory reading. Many first class draughtspeople will never have heard of them, but they are of interest in many ways. You will find them in reference libraries. Some are very expensive. They are worth looking at and, since they are reference books, you may look at them now and again all your lives. They are not in any order.

'Flatland: A Romance of Many Dimensions' by a Square (Edwin Abbott Abbott), published 1884, reprinted 1926, 1932 by Basil Blackwell, Oxford. (Did he call himself a square because his name was Abbott Abbott?)

Nature and Form, by D'Arcy Thompson, ed., Medawar.

Possible Worlds, by J. B. S. Haldane, Heinemann/Chatto and Windus, 1940.

History of Architecture on the Comparative Method, by Banister Fletcher, eighteenth edition, Athlone Press, 1975.

Ingenious Mechanisms for Designers and inventors, published by Industrial Press, 1946–67, New York.

The Planiverse: Computer Contact with a Two-Dimensional World, by Alexander Dewdney, Pan, 1984. (A development of the ideas in *Flatland*.)

Graphics is a very fast moving art form at the moment. New ideas and products keep appearing. Look at a trade paper such as *Reproduction* every few months, to find what is happening in the real world.

Index